RESEARCH TECHNIQUES
IN HUMAN ENGINEERING

RESEARCH TECHNIQUES IN

HUMAN ENGINEERING

ALPHONSE CHAPANIS

BALTIMORE | THE JOHNS HOPKINS PRESS

© 1959, The Johns Hopkins Press, Baltimore 18, Maryland

Distributed in Great Britain by Oxford University Press, London

Printed in the United States of America by The Horn-Shafer Company, Baltimore

Library of Congress catalog card number 59–10765

Second Printing, 1962

TO | **LINDA AND ROGER**

Preface

Human engineering is the name applied to that branch of modern technology which deals with ways of designing machines, operations, and work environments so that they match human capacities and limitations. Another way of saying this is that human engineering is concerned with the engineering of machines for human use and the engineering of human tasks for operating machines.

Whenever practitioners in this business get together one topic they are sure to discuss is their general dissatisfaction with the term *human engineering*. Some tangible signs of this malaise are the alternative titles which are proposed from time to time and used with varying degrees of acceptance. Some of these are *biomechanics, biotechnology, applied experimental psychology, engineering psychology, human factors in engineering,* and *ergonomics.* None of these substitutes, however, seems really to have taken hold. So, despite some uneasiness and residual uncertainty about what to call their discipline, most of the people in it answer to the name of *human engineer.* In using it in this book I believe that I am following current practice.

In a literal sense human engineering is as old as the machine itself. Machines have always been designed for human use, with more or less success. It was World War II that provided the impetus for setting off this area of technology as a separate discipline with a name of its own. Radar, aircraft, rockets, submarines, electronic computers, and hundreds of other instruments produced for the war were of a degree of complexity which the world had never seen before. More important, however, was the discovery that guesses, hunches, intuition, and common sense were no longer adequate for solving many problems of human use which were the result of this complexity.

vii

Indeed, in their efforts to match modern machines to their human counterparts, engineers found it necessary to seek the advice and counsel of many different kinds of biological scientists. That body of information we now have about man as a user of machines, or as an element in man-machine systems, constitutes the subject matter of human engineering.

Because man is himself such a complicated mechanism he can be studied from many different points of view. Human engineers draw upon many specialties for the information they need in their everyday work. Anthropometry (the measurement of body sizes), biology, physiology, toxicology, medicine, psychology, and industrial engineering all contribute relevant data and findings. The methods used by human engineers in collecting the information they need for the solution of their problems are correspondingly diverse. Some come from industrial engineering; others from the more exact sciences.

In 1954 a committee made up of representatives of the three military services (the Executive Council of the Joint Services Steering Committee for the Human Engineering Guide to Equipment Design) commissioned me to prepare a survey of research methods available to the human engineer. The result of that assignment was a monograph entitled *The Design and Conduct of Human Engineering Studies*, published in 1956. It had many shortcomings, more painfully evident to me than to my severest critics. Despite these inadequacies the response to the *Design and Conduct* was so much better than I had anticipated that I was encouraged to revise and expand it for commercial publication. The present book is based on that original monograph.

Support for the preparation of *The Design and Conduct of Human Engineering Studies* came from three different sources. Much of the writing was done while I was a consultant employed by the Operational Applications Laboratory of the Air Force Cambridge Research Center. To pay for the preparation of illustrations and the cost of typing the manuscript I drew upon the resources of my Contract N5-ori-166, Task Order I, Project Designation No. NR-145-089, between the Office of Naval Research and The Johns Hopkins University. The actual publication of the monograph was financed by Contract Nonr-1268(01), Project NR 145-075, between the Office of Naval Research and the San Diego State College Foundation. In so far as the present book stems from that original work I should like to acknowledge gratefully the financial and other support received from those three sources.

In rewriting and expanding the original publication into its present form I have again drawn upon the resources of Contract N5–ori–166, Task Order I, and Contract Nonr-248(55) between the Office of Naval Research and The Johns Hopkins University, especially for the preparation of new illustrations and for secretarial and typing assistance. Accordingly, it gives me pleasure to acknowledge this generous support of my activities by designating this book Report Number 1 under the latter contract. Reproduction of this report in whole or in part for any purpose of the United States Government is hereby permitted.

It also gives me pleasure to express my appreciation to John Wiley and Sons, Inc., for granting me permission to reproduce certain materials from their publications. Figures 1, 30, 34, and 35 are reprinted from Chapanis, A., Garner, W. R., and Morgan, C. T., *Applied Experimental Psychology: Human Factors in Engineering Design* (1949). The quotation on pages 13–14 is reprinted from Geldard, F. A., *The Human Senses* (1953); and the quotations on pages 11 and 12 from Jeffress, L. A. (ed.), *Cerebral Mechanisms in Behavior* (1951). To Prentice-Hall, Inc., I owe thanks for permission to reproduce Table 4 from Mundel, M. E., *Motion and Time Study: Principles and Practice*, 2nd Ed. (1955).

No book is the work of but a single author. To my students, colleagues, friends and critics I owe a special debt of gratitude for pointing out flaws, inaccuracies, and obscurities in the manuscript. Dr. Joseph Sidowski supplied me with an extensive set of valuable criticisms. Professor Wendell R. Garner, Dr. Natalia Potanin, and Mr. Leonard Horowitz read substantial portions of the text, giving me much useful advice. To these and many more my most heartfelt thanks. Those defects which still remain in this book are not for want of their trying but are due rather to my own obstinacy.

Finally, this book would never have reached the printer's shop without the patient and careful work of my laboratory assistant, Miss Lee Steinwald, who typed most of the first draft, and my secretary, Mrs. Dolores Pubanz, who typed successive drafts and the entire final copy. Their good-humored acceptance of my idiosyncrasies lightened the load of author and publisher alike.

Alphonse Chapanis

Baltimore, Maryland
December 23, 1958

Contents

| Chapter 1 | *Introduction* |

One important engineering lesson to come out of World War II is that "Machines do not fight alone." The war needed and produced many novel and complicated machines, but often these contrivances did not do what was expected of them because they exceeded or did not match the capabilities of their human operators. For example, radar has been called the eyes of the fleet. But radar does not see. It is a man who sees, and a piece of radar gear with many excellent features may be almost useless if the operator cannot see the information the radar presents him on the scope. Similarly, a fighter aircraft is a failure if a man cannot fly it safely and effectively. A guided missile will reach its target only if the men who check, fuel, and service it can do their jobs without error. And so we could go through the whole catalog of modern machines of war. This realization led to the rapid development of the field of *human engineering*—a branch of applied science aimed at matching machines and tasks with the abilities of their human operators.

After World War II human engineers turned their attention to the thousands of machines which surround us in our workaday world. There they found that many of the same design errors which plagued the sailor, soldier, and airman exist in factories, tractors, trucks, automobiles, and even in the home. Gages which the industrial worker can scarcely interpret, controls which mystify the housewife, highway signs that confuse the motorist—these and thousands of other similar examples are all instances of poor design. Moreover, the petty annoyances, the irritations, the inefficiencies, and the accidents caused by these design errors multiply daily as our civilization becomes more and more mechanized. And so it is that, added to the vast number of

1

factors they have to keep in mind when they design equipment, military and industrial engineers are now being forcefully reminded of the importance of that nebulous one called the human factor.

To do his job well today the design engineer needs to have data and measurements on all kinds of human capacities, abilities, and limitations. Can people read green reflectorized signs more easily than blue ones? Are telephone numbers which make use of an exchange name and number, like STate 9–9722, easier to remember than the equivalent set of digits, that is, 789–9722? Is a number like 789–9722 easier to remember than 7–89–97–22? How much aiding should be provided in power-assisted steering wheels? What is the most comfortable height for an ironing board? Are push-button controls on a stove better than knobs? Should switches move to the right or left to turn something on?

It would be ideal if the engineer could turn to handbooks of human data, or to textbooks of psychology, physiology, or anthropometry to find the information he needs, in about the same way that he can turn to the *Handbook of Chemistry and Physics*, the *Radio Engineer's Handbook*, or any of a dozen others. Although there are a few books which try to present human data in this way[29, 78, 107, 117] * our storehouse of information about the human factors in engineering design is still too limited to provide more than a small percentage of the answers we would like to have. In lieu of direct help, then, many engineers have turned to a procedure they know well—the procedure of "trying it out." The trouble with this is that trying something out on people is a lot more complicated than appears at first glance. Let's take a quick look at some reasons.

Some difficulties of human experimentation. When you do a chemical experiment, you can reach up onto your shelf for a reagent and look at the label to see what you have. Most likely the label will identify the reagent by name, tell you its purity, and give you its chemical formula. In addition, for a large variety of chemicals, you can be sure that it will be the same this afternoon as it is this morning and that it will be the same tomorrow morning as it is today. You can say anything you want to it, and it will still be what the label says.

But contrast this situation with what you get when you reach out into the corridor for a human "reagent." You literally do not know what you have. At best, you have a little information about the past history of your human reagent, some rough idea about the limits of its

---

* Numbers refer to items in the bibliography starting on page 303.

performance, and only a vague notion about its stability. You know for sure that it will not be the same this afternoon as it is this morning and that it will definitely change overnight. Finally, you can be certain that its reactions will be markedly determined by what you say. Complications like these are the reason for this book.

THE AIM OF THIS BOOK

The aim of this book is twofold. Its first purpose is to describe some of the methods available to the human engineer for collecting trustworthy data on men and machines and the relationships between them. Its second purpose is to discuss some principles and guide lines about ways of doing dependable studies on people.

We shall not be concerned here with abstract philosophical discussions about science and scientific method. There are excellent books on the philosophy of science and experimental inference,[35, 37] and the reader who is interested in this approach to the problem should consult them. There are also a number of books which discuss scientific research in general terms—Wilson's text being a good one in this field.[116] Unfortunately, books such as Wilson's are generally so broad in coverage that they cannot get to the special kinds of problems one finds in studying people. At the other extreme, textbooks such as Andrews,[3] Brown and Ghiselli,[17] O'Neil,[87] or Townsend[106] contain a wealth of information about a great variety of basic kinds of psychological experimentation which the practical man will never encounter.

This book, then, is practical and selective—being oriented toward those principles of human experimentation which are primarily useful in the study of men and machines. In addition, numerous examples of both good and bad studies are discussed here since I have relied heavily on the method of example as the best way of explaining principles.

CAN SCIENTIFIC METHOD BE TAUGHT?

If someone offered to teach you "How to Become an Electronic Engineer" in one easy lesson, you would probably regard him as a charlatan, or completely out of touch with reality. You would undoubtedly feel the same about anyone who offered to tell you "How to Become a Good Research Man" just by reading a single book. This one has no such pretentions.

Research is unique and hard to characterize. Science has been defined as "doing one's damnedest with one's mind, no holds barred."[14] There is a lot of truth in this statement. Methodologies in the various sciences undergo continual change, and the mark of a successful investigator is that he can find novel methods of solving his problems. Another way of saying about the same thing is this: research is not a repetitive production process. No two studies are exactly alike. If they were, the second would not be worth doing. This unique character of scientific investigation makes it hard to find rules which define or characterize the process we call research.

Human engineering research is broad in scope. Another difficulty is that human engineering studies cover such a wide spectrum. They extend over the whole range of specificity from those which attempt only to answer a particular question to those which try to arrive at answers applicable to a whole class of situations. Among the former we find studies aimed only at answers to questions like these: "Is dial A more readable than dial B?"; or "Is this warning bell more audible than that?" Among the latter, we may be interested in the answers to such sweeping questions as: "Is it possible to fly an aircraft in which the primary flight instruments are replaced with complex sounds?" or "In an air traffic control tower what human functions should be replaced by automatic machine operations?" Finding principles common to the methods suitable for such diverse kinds of studies is difficult indeed.

In research, as elsewhere, experience is a good teacher. Then, too, there is no doubt that much of the skill which accomplished investigators demonstrate in their work has come from actual experience. The mature research man has learned what will and will not work from trying various things and finding out which are successful. In a very real sense it is impossible to replace this kind of experience. That is one reason why college students spend so much time getting the feel of "science" in laboratory exercises and why the research thesis is such an important requirement for the M.S. and Ph.D. degrees.

Some tactics and strategy of research can be learned. Arguments such as those above make some people skeptical about the possibility of ever being able to teach scientific methodology. In this book, we shall not adhere completely to this pessimistic point of view. The fact of the matter is that some studies are untrustworthy for reasons you can identify and put into words. Moreover, the same mistakes in technique and methodology keep appearing and reappearing. For these reasons, then, we shall accept as a major premise that research methodology cannot be learned completely from a book. But we shall also accept as a premise that some aspects of the tactics and strategy of science can be put down in the form of general rules and principles. These principles can be taught, and they serve as important guideposts and warning signs to mark the path of the investigator who attempts to study that most complicated of all mechanisms—man.

ON COMMON SENSE VERSUS RESEARCH

One of the most common criticisms which the behavioral scientist hears from physical scientists and engineers is: "Why all this talk about research and experiments? After all, human engineering is nothing more than common sense." There is both fact and fallacy in this statement. Many "human engineering" principles do, in fact, involve little more than "common sense." The idea that seats in automobiles ought to be built to accommodate the range of sizes drivers come in—this is a "common sense" idea. You do not need research to tell you that the headband on a telephone operator's headset should not be so tight that it hurts her ear. Nor is any experimentation required to justify the conclusion that the dials and gages on machines should be where operators can see them or that controls should be where operators can reach them. Many human engineering improvements can be made by any intelligent person who is aware of these problems and is willing to approach them primarily in terms of the operator.

Then, too, many of our present-day machines are actually very good, even though they may not have been designed on the basis of human engineering experiments. The modern telephone handset is a very convenient and well "human engineered" instrument as compared with its primitive ancestor of sixty or seventy years ago. Yet little in the way of scientific experimentation on people was done to bring about these changes.

Modern computing machines, typewriters, automobiles, stoves, irons, and washing machines are all much handier gadgets than they were twenty-five years ago. Human engineering research points to a number of ways in which some of these devices can be improved still further, but for the moment our point is that substantial improvements have been made in all of them without the benefits of careful *human* research.

Common sense ideas change. There are, however, some very good reasons why "common sense" cannot always be trusted. The first is that our ideas of what is, and is not, common-sensical are undergoing continual revision. Moreover, the changes in our ideas of what is common-sensical are frequently the results of research. Things that we take for granted today were not so obvious a short while ago. Just a little more than fifty years ago common sense told everyone—that is, nearly everyone—that women would never make good stenographers, secretaries, and office workers.

From our vantage point today it is almost impossible to believe the jokes, editorial jeers, and ridicule which the nation's newspapers used in describing Samuel P. Langley's attempts to construct a flying machine in 1901. The whole idea was ridiculous—it violated all reason and common sense! The eminent mathematician and astronomer, Simon Newcomb, wrote an article calmly pointing out the scientific reasons why such a venture was doomed to failure. Rear Admiral Melville, then Engineer-in-Chief of the United States Navy, wrote:

> Outside of the proven impossible, there probably can be found . . . no field where so much inventive seed has been sown with so little return as in the attempts of man to fly successfully through the air. . . . It may be truly said that, so far as the hope of a commercial solution of the problem is concerned, man is today no nearer fulfillment than he was ages ago when he first dreamed of flying through the air. . . . A calm survey of certain natural phenomena leads the engineer to pronounce all confident prophecies at this time for future success as wholly unwarranted, if not absurd. (Melville,[82] pp. 820–22)

All this was in 1901—two years before the Wright brothers' successful flight at Kitty Hawk!

Look at what L. V. Berkner, President of Associated Universities, Inc., wrote recently to a Congressional committee investigating military research and development:

> The (Office of Scientific Research and Development) files are replete with statements of high military authority concerning these 'ridiculous' ideas. . . . Early in the war, the idea of night fighters operating from carriers was met with actual derision. But, as soon as experimental night fighters had been built and operated from carriers against actual targets, all resistance to their employment disappeared. Thereafter, night fighters could never be procured fast

enough to meet the enthusiastic requirements of the Navy. In this present day of complete acceptance of carrier-based night fighters, it is difficult to realize the resistance which was generated to such an obvious weapon during its development.[65]

This history is particularly relevant to our story because human engineering studies of dark adaptation, red goggles for dark-adapting the eyes, red lighting for instrument panels, and related problems, contributed in some small measure to the successful marriage of man and machine in the employment of this weapon. More important, however, is the moral it holds about the fickleness of "common sense."

Common sense ideas are often wrong. A second and more important point is that even today common sense can lead you astray in the solution of many human engineering problems. Put the following problem to the average person and see what answer you get. Nearly everyone knows that boiler factories are very noisy and that workmen in boiler factories have difficulty hearing and understanding speech because of this noise. Does it seem sensible that you can improve the intelligibility of speech by putting earplugs in the workmen's ears? The chances are very high that the answer you will get is that this is a nonsensical notion. Perhaps even *you* think so. If you do, research proves you wrong: Workmen can hear better with earplugs! For an explanation see Chapanis, *et al.* (pp. 223–24).[29]

Here is another example. In the cockpits of many modern aircraft there is an instrument called an artificial horizon (see Figure 1). In the conventional instrument a miniature representation of an airplane remains stationary and a white line, traversing the indicator, represents the horizon. During straight and level flight the horizon line is flat. When the aircraft banks to the right, the horizon line rotates counterclockwise, i.e., to the left. In the upper right-hand illustration in Figure 1, the horizon line extends from 7:30 to 1:30 o'clock. When the aircraft banks to the left, the horizon line rotates clockwise. The person who designed this indicator undoubtedly relied on his common sense for this arrangement, and there is a logical argument to support his decision. After all, when you sit in an aircraft cockpit and the plane banks to the right, the horizon does appear to tilt to the left. Hence, if you construct an instrument which duplicates this picture you should have an instrument which is easily and readily interpreted. Seems sensible, doesn't it?

Yet the results of research show this kind of a display to be confusing, even to experienced pilots. A much more easily interpreted indicator is

one in which *the movement relations are exactly reversed*. Now that we
have the results of this experiment, it is not hard to find a common
sense explanation for it: The pilot thinks of the earth and horizon as
stable references with the plane moving around in space. An instru-

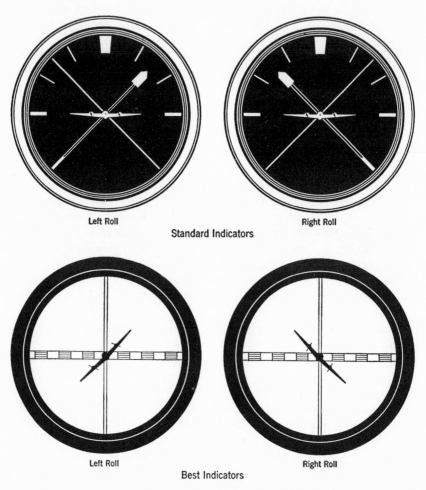

FIG. 1. Two kinds of artificial horizon showing a left roll and right roll. The two
top indicators have been standard in aircraft. The lower indicators have movement
relationships which are the reverse of the top ones. Research shows that the bottom
ones are more easily interpreted than the top ones. (Data of Loucks[74] from Chapanis,
Garner, and Morgan.[29])

ment which shows these relationships directly is the more easily interpreted.

Try some common sense on this problem. To conclude this discussion let's look at just one more example. Many kitchen stoves are designed something like those illustrated in Figure 2. There are four burners (or electric coils) arranged in a square on the top surface of the stove. The four controls for the heating units are aligned on the front panel of the stove at right angles to the top surface. One of my former students, Lionel Lindenbaum, became interested in the linkages between the controls and burners. A survey of existing stoves showed that they are being manufactured with many different kinds of control-burner

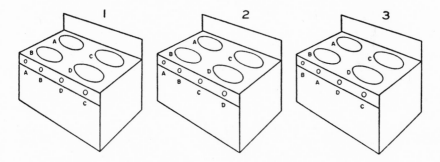

FIG. 2. These are three different control-burner linkages commonly found on stoves. Which do you think is best? (After Chapanis.[28])

linkages, only three of which are shown in Figure 2. With this as background, then, Lindenbaum designed an experiment to answer two questions:

(1) Does the control-burner linkage make any practical difference in the time and errors made in operating the correct control for a given burner?

(2) If there are any differences among these arrangements, which one is best after a long series of training trials?

For his experiment Lindenbaum used a wooden model of a stove with paper rings to represent the burner units. In the center of each burner unit were two small lights. One of these could be turned on by the experimenter; the other by operating the appropriate control. The subject's task was to operate the control which would turn on the light immediately under the experimenter's light. The experimenter turned on the various lights in a random order. The subject started each trial with his thumb on a starting button centered in front of the stove. Fifteen subjects were tested on each stove and each subject was given 80 consecutive trials. The control-burner linkages were demonstrated and explained before the first trial.

Although a number of other technical details have been omitted, this description should be sufficient for our purposes. What does your common sense tell you about the outcome of this experiment? What are your answers to the two questions Lindenbaum asked himself? The probability of your getting the answers correct by chance is rather high—one in four—because the possible alternatives are these:

Alternative	Question 1	Question 2
1	No	—
2	Yes	1
3	Yes	2
4	Yes	3

For this reason you should try these questions on a sample of 10 people at least. In a sample this size you can generally find at least one person who is willing to defend each of the four alternatives above. How much help is common sense in a case like this? For the correct answer, see the end of the chapter.

In summary: Be careful of common sense. To sum up this section, then, common sense is too shifty a standard upon which to base design decisions. A science of human engineering built on common sense is like a house built on quicksand. The hard core of reliable information in this field must come from careful studies and experiments. This is the only firm foundation we can trust. The fact that you, your father, and your grandfather have always done something in a particular way does not prove that this is the best way of doing it.

In the field of human engineering do not be misled by the fact that a machine works. If you are not careful you can, in fact, get into a curious circular argument. Take dental equipment: Dentists say there cannot be very much wrong with their equipment since they buy it from the best firms. The manufacturers say there cannot be very much wrong with their equipment since the best dentists buy it. Neither is any guarantee that it is designed as well as it might be.

A GENERAL PHILOSOPHY ABOUT
HUMAN RESEARCH

Before we get any deeper into the subject matter of this book, it would be well for us to look at some general points concerning the matter of research on people. These apply to almost everything we shall have to say later, and, in a manner of speaking, we may say that they help to establish a basic point of view, or philosophy, about this subject matter.

SOME DIFFICULTIES IN STUDYING PEOPLE

Turn to the first chapter in almost any textbook of elementary psychology, and you are almost certain to find there a discussion of scientific methodology. On the other hand, you will rarely find a comparable chapter in textbooks of elementary physics or chemistry. Why is it that psychologists are so concerned about methodology? The answer is that, in some respects, doing research on people is the hardest kind of experimentation there is. There are a number of reasons for this.

Man is a complex organism. The first and most important reason is that man is physically very complicated. In fact, there are some people who are willing to argue that he is more complicated than any other physical system in existence. Modern computing machines which add, subtract, and do even some rudimentary kinds of logical operations with fantastic speeds make newspaper headlines almost daily. Yet compared with the human body, even the most complex computing machine is a child's toy. Here is what the eminent mathematician, John von Neumann, had to say about the subject:

> In comparing living organisms, and, in particular, that most complicated organism, the human central nervous system, with artificial automata, the following limitation should be kept in mind. The natural systems are of enormous complexity. . . . With any reasonable definition of what constitutes an element, the natural organisms are very highly complex aggregations of these elements. The number of cells in the human body is somewhere of the general order of 10^{15} or 10^{16}. The number of neurons in the central nervous system is somewhere of the order of 10^{10}. We have absolutely no past experience with systems of this degree of complexity. All artificial

automata made by man have numbers of parts which by any comparably schematic count are of the order of 10^3 to 10^6. In addition, those artificial systems which function with that type of logical flexibility and autonomy that we find in the natural organisms do not lie at the peak of this scale. The prototypes for these systems are the modern computing machines, and here a reasonable definition of what constitutes an element will lead to counts of a few times 10^3 or 10^4 elements. (von Neumann,[111] pp. 2–3)

In the discussion which followed the paper from which the above quotation was taken, the psychiatrist Warren McCulloch had this to say:

I confess that there is nothing I envy Dr. von Neumann more than the fact that the machines with which he has to cope are those for which he has, from the beginning, a blueprint of what the machine is supposed to do and how it is supposed to do it. Unfortunately for us in the biological sciences—or, at least, in psychiatry—we are presented with an alien, or enemy's machine. We do not know exactly what the machine is supposed to do and certainly we have no blueprint of it. (von Neumann,[111] p. 32)

The point of citing these men at such length is to emphasize from the start that when you work with people you are working with enormously intricate organisms whose internal workings are imperfectly understood.

The scientific study of man has a short history. The second reason why it is hard to do good experiments on people is easy to understand. Whereas chemistry and physics have been studied for hundreds of years, experimental psychology, as a science, is less than a hundred years old. Indeed, there are even some people today who question whether it is at all possible to subject men to scientific scrutiny. However, we shall ignore this minority, and go along in the belief that it is not only possible but is being done every day.

The short history means, however, that we have only begun to explore ways of doing research on people. We still have much to learn, and there are many human problems which cannot be tackled, simply for want of valid techniques.

Man is not a good observer of complex events. A third reason why it is hard to do good experiments on people is that man is not a very good observer—especially when it involves other people. Many courses in elementary psychology start out their laboratory exercises with a convincing demonstration of this point. One such demonstration is to show the class a short, carefully-staged motion picture film of some dramatic event, for example, a robbery. Immediately after viewing the film, the audience is asked some simple questions about it—things like:

1. Were the window shades up or down?
2. What time of day did the event occur?
3. Was the assailant a man or a woman?
4. In which hand did the assailant hold the weapon?

Even when the viewers are forewarned that (a) they are to see a movie about a crime and (b) they are to answer questions about it later, the accuracy of the observations is astonishingly low. One difficulty is that the observers become *ego-involved* (as some psychologists might put it) in the action they witness. It is very hard for them to remain aloof from the scene and be objective recorders of the events that transpire. Another problem is that the average person has, as a psychologist might put it, a very narrow *range of attention*. It is impossible for a person to pay attention to very many things at the same time. Some of the practical consequences of this will be discussed later when we come to the section on observation.

Finally, we must also beware of illusions and differences in sensory capacity which may produce discrepant observations. For example, about one man in every sixteen is color blind to some extent. A famous controversy in the history of astronomy has been traced to the fact that one of the scientists involved was color blind. Tone deafness, taste blindness, and other related types of defects occur often enough to constitute a problem whenever human beings act as observers.

Man does not think of himself objectively. The next reason why it is hard to do good experiments on people is related to the one above. It is well stated in the following quotation from Geldard:

> There is probably no topic in connection with which the average layman is more ready to pronounce a judgment or express an opinion than that of psychology. We may be content to leave the intricacies of physics or chemistry to experts in these fields, satisfying ourselves with a few basic ideas about the operation of the complicated world around us. But the business of living and conducting ourselves in a world of human beings is very much the concern of each of us. We are all under the necessity of acquiring and pressing into service a workable set of conceptions of human nature. Moreover, the materials with which psychology attempts to come to grips are immediately available in the form of our own thoughts and feelings and in the actions of others. We come readily to generalize from our experiences and to develop a set of beliefs concerning the operation of the human mind. The demand for working principles is so insistent that it is not surprising that hasty convictions, half-truths,

even superstitions become lodged in our mental constitutions and sometimes are modified or expelled only with the greatest difficulty. We feel "from experience" that we are rather good judges of the motives of other people. We have "explanations" for the fact that some persons have good memories while others have not. We acquire beliefs as to the relative influence of nature and nurture in shaping our behavior. We do not hesitate to ascribe the genesis of anomalies we observe in others to temperament and character "types." We have convictions concerning the effects of sleep, the importance of dreams, the influences of age, sex, fatigue, climatic conditions, and a host of other matters. In fact, unless we turn the searchlight of self-criticism upon these beliefs we may go on indefinitely, trusting our crude observations, never pausing to draw into question the processes whereby we form our prejudices and set up our standards. (Geldard,[59] pp. 1–2)

In order to do good studies on people, it is necessary to counteract the interference produced by the experimenter's idiosyncrasies and prejudices. To some extent, we see what we want to see or expect to see. If you really believe that the study of mathematics gives you a logical mind, you will have no trouble finding cases to support your point of view, and you will probably find it hard to think of a case which is contrary to your beliefs.

The history of science contains many examples of obvious data being overlooked by a researcher because of his hypotheses, or theories. Objectivity is hardest to achieve in the relatively undeveloped fields of scientific inquiry—and the study of man certainly falls in this category. It is at this point, too, that we come face to face with the near circularity of scientific method: The elimination of bias depends on an extensive body of reliable information; the acquisition of reliable information requires freedom from bias. Fortunately for us, the progress of science is an ascending spiral rather than a closed circle.

Some human experiments are dangerous to do. Research on people is also limited by the fact that there are certain kinds of experiments you cannot do because they are too dangerous. The engineer is frequently able to learn important things by "testing to destruction." The analogy holds in human experimentation too, except that these experiments are impossible to do. How much deceleration can the human body tolerate? The answer to this question has some important implications for the design of safety equipment in aircraft, automobiles, and other vehicles. Or, how much overload can you put on a radar network before the human components break down? Here again the answer to this question is important, but it is a question which is very difficult to answer.

We should note that the problem comes not only from the reluctance of experimenters to carry out such studies but also from the difficulty of getting subjects to participate in them. In stress experiments, your subjects may walk out of the experiment when the going gets tough. The problem of enlisting full, wholehearted co-operation in such studies may be very difficult indeed.

Some experiences are not accessible for study. The final obstacle to the study of the human organism is that a large number of personal experiences are not open to inspection. Science is a public affair, but things like *consciousness, ideas, mental images, fatigue,* and *motives* are highly personal experiences which cannot be looked at, handled, counted, or measured. There is no way the psychologist can get into your head to see what it is you see when you say, "I see red." For all I know perhaps what you see and identify as red is the sensation which I call green.

The study of motives—the reasons why people do what they do—is particularly treacherous. If you ask people why they like this car, prefer that chair, bought this television set rather than that one, or made that mistake in driving, they will probably give you an answer. The only trouble is that the answer you get is almost certainly not the correct one for human engineering purposes. Probing for these hidden mainsprings of human action is a difficult and time-consuming business that is best left to the psychiatrist or clinical psychologist.

A GENERAL STRATEGY

The preceding section listed some of the general reasons why it is hard to do experiments on people; later we shall get down to some more specific matters. The general tenor of the section above is negative since it emphasizes the difficulties of doing human research. But there is a positive side to the story as well. Let us look at the points, one by one, and see what lessons they contain:

Do not underestimate man. Many physical scientists seem to approach human experimentation with the attitude, "Well, after all, man is nothing more than. . . ." You can add your own ending to the sentence, since the words have changed over the years and will continue to change. A couple of hundred years ago it was the vogue to say that man is nothing more than a system of complicated levers and pneumatic tubes (the nerves and blood vessels) which carry energizing liquids. Fifteen or so years ago, it was popular to say that man is nothing but a servo. Now it is the thing to say that man is nothing but an information-handling channel. Call man a machine if you will, but do not underestimate him when you experiment on him. He's a non-linear machine; a machine that's programmed with a tape you cannot find; a machine that continually changes its programming without telling you; a machine that seems to be especially subject to the per-turbations of random noise; a machine that thinks, has attitudes, and emotions; a machine that may try to deceive you in your attempts to find out what makes him function, an effort in which, unfortunately, he is sometimes successful.

Be alert to new techniques. We do not have adequate methods for finding out all the things we need to know about people. Above all, we need novel and imaginative techniques for the study of man. This is an area in which behavioral scientists can learn much from the engineering and physical sciences.

Although concepts like feedback, feed-forward loops, and the like have not solved all the problems of human behavior, there can be no doubt that the servo-system analogy has been a very useful one in studying certain man-machine problems. Similarly, communication theory has given us new statistical methods which have been highly productive in studying certain other kinds of problems. The be-havioral scientist owes a real debt to the physical sciences and to engineering for developments in his own field. For this reason, the investigator who undertakes to study man will do well to keep his eyes open for still other applications of this sort.

Remember that no book can ever supply the most important in-gredients a good research man needs—imagination, adaptability, and inventiveness.

Set up conditions of observation with great care. In so far as possible, the data of human experiments should come out in an objec-tive, permanent record that can be studied and gone over later. If it is not feasible to get permanent records, make it as easy as possible for the observer to record immediately what he observes. Define carefully, in advance, what the observer is to record and work out a simple short-hand notation he can use in recording his observations. The time-and-motion symbols (described in the next chapter) are one such convenient system. Other specialized kinds of notation can be devised to suit particular purposes. One example is the motion-time code devised by scientists at the Systems Research Laboratory at Johns Hopkins for recording the activities of a radar operator and plotter in shipboard combat information center (CIC) operations (see SRL Research Report No. 3,' pp. 9–10).

Standardized check lists, record sheets, and data forms also serve a useful function in helping the observer record his data with minimum inconvenience. Although it might seem a needless waste of time to do so, preparing a special data form for each experiment or series of observations is usually well worth the effort. The cost of duplicating a few dozen or a few hundred data sheets is trivial, and you have a data sheet tailor-made for each special set of observations. Moreover, the blanks on such a data sheet serve as a constant reminder of the items the observer should be recording—things like the date, time of day, observer's name, conditions of observation, and so on. A good rule to observe in this connection is to write down more information than you think is essential. Don't trust your memory to recall that the subject remarked he had lost a lot of sleep the previous night, or that the equipment needed adjusting in the middle of an experimental session. Often you will find such notes of great help in explaining some aberration in the data or in reconstructing the conditions under which a series of observations took place. If you record too much of this supplementary information, you can always ignore it later. But if you did not record such supplementary information on the spot, you cannot recapture it later.

Make the observer responsible for observing a very small part of the action that transpires, and, if necessary, have other aspects of the observing done by still other observers. Try to cultivate attitudes of objectivity in your observers; have them then remain as aloof from the action as possible. If at all possible, have more than one observer record the same action. Psychological experiments show that the pooled statements of several observers are uniformly more valid than those of a single person.

Cultivate attitudes of objectivity in thinking about people. In your experiments on people, remember that your biases, prejudices, and hunches may well be wrong, and they will undoubtedly color what you see. Be as objective as you can. Be on the lookout for negative instances, for data that do not agree with your ideas, and for possible flaws in your technique. Become, in other words, a confirmed skeptic. In particular, be alert to the possibility that your "common sense" ideas may be wrong.

Perhaps the most important characteristic of scientific observation is that the scientist observes and records *facts*. He evaluates the facts later. Popular observation very often confuses evaluations with facts. If a worker in an office is sitting with his feet up on his desk, this is an observation which the scientist records as such. He does not make the mistake of recording, as an office manager might, that the worker was loafing. The worker might or might not have been loafing. One way of increasing the objectivity of recordings is to decide, in advance, on criteria for observation, that is, exactly what kind of behavior is going to be recorded and how. Examples of such criteria are given on pages 30 and 31 in Chapter 2.

Three approaches to dangerous experimentation. When experiments are potentially dangerous to the human beings who might participate in them there are three ways to proceed:

1. Use animal subjects and extrapolate from their behavior to that of humans. As an example, many high-acceleration experiments and studies of explosive decompression have been done using chimpanzees monkeys, and other animals. Although they are expensive, chimpanzees and monkeys are often used for this purpose since they resemble man so closely in certain important anatomical and physiological respects. Autopsies on the animals then permit the medical experimenter to extrapolate to the human case.

2. Use dummies—either scale models or full-size models. Studies of pilot ejection seats and parachute opening devices were first done with dummies having roughly the same physical characteristics as the human body. One aircraft company conducted tests on the aerodynamic properties (spin and tumbling) of certain pilot ejection seats using models and speeds which were scaled down to simulate the characteristics of the human body in highspeed flight. After these initial tests, experiments were then carried out with full-scale models ejected from rocket sleds. Studies of crash injuries in aircraft and automobiles are commonly studied in this way.

3. Use human volunteers who understand fully the risks they are taking. This approach often follows initial experiments of one or both types mentioned above. Generally, tests are started at safe values and breakdown conditions are approached in successive small steps. In this way, the experimenter can often get a good idea of the maximum stress a man can take before any permanent harm is done.

Operationism and the study of behavior. In several places throughout this book we shall see that some ways of looking at human beings are so vague as to be useless, while others are meaningful and researchable.

The general rule is that we can study human activities only by studying *behavior*, the things people *do*. We cannot study what they experience. Since this principle is so important, let us examine it in greater detail, first by way of a bit of history, and then in relation to our subject matter.

In 1927 the physicist Bridgman introduced an important new idea into scientific methodology. It is that a scientific concept can be defined only in terms of the operations which are necessary to measure it. Let's look at his own words:

> The new attitude toward a concept is entirely different. We may illustrate by considering the concept of length: what do we mean by the length of an object? We evidently know what we mean by length if we can tell what the length of any and every object is, and for the physicist nothing more is required. To find the length of an object, we have to perform certain physical operations. The concept of length is therefore fixed when the operations by which length is measured are fixed: that is, the concept of length involves as much as and nothing more than the set of operations by which length is determined. In general, we mean by any concept nothing more than a set of operations; *the concept is synonymous with the corresponding set of operations.* (Bridgman,[15] p. 5)

To see how this principle is put into practice note how the author of an elementary textbook in physics starts his discussion of force: "Force is a concept which can be derived in terms of the fundamental concepts of length, time, and mass." In other words, force is what you measure in this way. Similarly, you can talk about electricity as a form of energy, but it can be studied only when you are told that electricity is measured in terms of certain effects it produces—light, heat, magnetic effects, or chemical changes.

The operational approach to the definition of concepts is well established in the physical sciences, either implicitly or explicitly. But when we get into problems of human behavior we tend to forget about measuring sticks, scales, and instruments. A while ago a group of engineers was discussing the feasibility of some research on human factors in the design of air traffic control centers. "What we really need," said one, "is some research which will tell us what goes on in the air-traffic-controller's mind." A psychologist on the same team started out, in all seriousness, on an investigation of the "mental images" which controllers form of the air space around the control tower.

Although we use terms like *mind* and *mental image* in our every day speech, for scientific purposes they are of no use whatsoever. As they are phrased, the questions are not researchable primarily because the concepts involved are not defined operationally. *Mind* is not a *thing* you can pick up, look at, or pass around. Although we may all agree that we have *mental images*, we cannot do any research on them until we agree on how to measure them.

To put this into concrete terms, consider the problem of color blindness. You could, if you wanted, talk about color blindness in terms of *mental images* or *sensations*, but the people who have tried this approach have not succeeded in finding out very much about the defect. This does not mean we are doomed to failure. Far from it. What we can do is to study the kinds of confusions color-blind people make. This tells us a great deal both theoretically and practically. For example, one of the principal forms of partial color blindness is called protanopia. The protanope:

1. Can match all the colors of the spectrum with mixtures of two suitably chosen primary colors, colors which most people call green and blue.

2. Will confuse a blue-green color (a wave length of about 495 millimicrons) with white or gray.

3. Will confuse colors which the normal person calls red, orange, yellow, yellow-green, and green when suitable brightnesses and saturations of these colors are used.

4. Has reduced visibility in the red end of the spectrum; that is, he cannot see light from the long wave-length end of the spectrum, which is easily visible to most people.

While there is much more that can be said about this type of color blindness, it is enough here to observe the way in which these statements are phrased. Notice that none of them says anything about how these *look* to the color-blind person. We do not know that. What we do know is that such a person makes color confusions which the normal person never makes. The patterns of these confusions can be studied because you can measure them and observe them. In fact, the experimental subject does not even need to speak. Every one of the statements above could be verified with animals by observing (in properly designed experiments) what the animals do. As an exercise try to imagine how this could be done.

The point of this discussion, then, is that we should be sure the psychological concepts we propose to investigate are definable, and by *definable* we mean *measurable*. If a concept cannot be defined in terms of some set of measurements, we had better examine our ideas about that concept again. Try your hand, for example, at making up operational definitions of *fatigue, carelessness,* and *information-handling capacity of the human channel.*

A PREVIEW OF THE CONTENTS
OF THIS BOOK

Having thus introduced the problem and set the general tone for our considerations of methodology, we may now ask: What are the methods we have at our disposal for the study of human engineering problems? If you look through the literature on the subject you will find that many ingenious ways have been used for getting good answers to practical questions. The techniques and methodologies are so varied that it seems, at first glance, almost impossible to categorize them. Nonetheless, they do fall into certain more or less clearly defined classes.

Chapter 2 is concerned with methods of direct observation. These are characterized by the fact that the researcher goes into an actual work situation and observes what goes on there. Chapter 3 deals with methods for the study of accidents and near-accidents. The material on statistics in Chapter 4 is of a general nature applying to all of the material in the book. Chapter 5 takes up the experimental methods, while Chapter 6 deals with some general problems involved in doing experiments on people. Chapter 7 is a study of a special class of experimental methods—the psychophysical methods—which are highly useful in certain types of human engineering work. Chapter 8 on articulation test methods is also directed toward a special class of methods used primarily in studying speech-communication systems.

ON THE EXAMPLES IN THIS BOOK

You have already had an introduction to something you will find throughout this book, namely, a liberal use of practical examples. In this chapter, for example, I have mentioned radar, aircraft, artificial horizons, earplugs, and stoves, among other things. In the remainder of the book other highly technical and complicated machines will be used to illustrate various principles. These are all real problems and I have deliberately picked a wide variety of them for two reasons.

To illustrate the scope of human engineering. By selecting diverse examples I have tried to illustrate the scope of human engineering and the many different kinds of problems which the practical man may be asked to tackle. These examples show what human engineers do and they show how the methods discussed in this book are actually used in their work.

Examples to suit the reader. I anticipate that many different kinds of people will read this book and that their backgrounds will be different. By selecting a wide range of examples I hope that almost anyone who reads this book will find some problems which seem familiar to him. Inevitably, of course, some of the examples will seem strange and unfamiliar.

Some advice to the student. Students who have been exposed to this kind of material are sometimes dismayed by the technical details of the machine systems which are discussed here. Thus, a word to the student is in order. First, remember that these are real examples; they are what human engineers concern themselves with. If you are interested in human engineering as a profession, you should be prepared to put some effort into learning something about the kinds of machines human engineers work on. Second, although some of these examples may seem very complicated, in actual fact they are all rather simple. I have deliberately avoided discussing some of the really difficult kinds of experiments human engineers get involved in. Finally, remember that this is a book on research techniques. It is usually not necessary to understand fully the particular machine to see the general principle about methodology which lies behind each example. I have used real problems for a reason, but you should also try to see through the example to the method. Then try to make up your own problems for each of the methods. By keeping these rules in mind you should be able to extract the maximum amount of value from this material.

ADDENDUM

The correct answer to the problem posed on page 10 is alternative 2.

Chapter 2 | *Methods of direct observation*

One of the most direct ways of attacking problems in human engineering is to go into an actual work situation and observe what goes on. The engineer can sometimes see ways of improving things, once he has an accurate and concise description of the present method of doing a job. Even when such a description does not actually suggest ways of making improvements, it often pinpoints important trouble spots or difficulties in an operation. This at least provides the engineer with a starting point by telling him where there are some gains to be made.

Time-and-motion engineers have been in the business of studying jobs for a long time, and, in the course of their work, have come up with some useful techniques. The emphasis in time study is on the determination of the "standard times" required to do various jobs as an aid to planning production and estimating costs. Many industries today also use time-study methods in setting up pay schedules for different jobs. These particular methods are of little interest to the human engineer and we shall not consider them here. Motion study, unlike time study, is concerned primarily with the movements human operators make in doing a job. Its aim is to find the best way of doing a job or ways in which a job can be simplified. It is this class of methods which are of direct interest to the human engineer.

Although there are a number of different motion-study methods discussed in this chapter, the emphasis is a little different from what you will find in a typical textbook of time-and-motion study. In addition, we shall discuss here several techniques which are usually not found in such textbooks. Perhaps the best way of describing the methods in this chapter is to say that the feature common to them all is that observations are made on genuine operating man-machine systems. Their aim is to generate ideas about ways of improving the job or the machines with which men must work.

OPERATOR OPINIONS

The person who is most familiar with a system and so, theoretically at least, most qualified to describe it is the man who uses it and works with it. Many industries capitalize on this storehouse of experience through the medium of the employee's suggestion box. Worker or user opinions can sometimes provide valuable suggestions for human engineering improvements or suggest avenues in which further research may be profitable.

A study of air traffic control problems. A good illustration of this technique appears in a report by Fitts,[49] who used it as a basis for studying air traffic control systems. Interviewers visited eleven Air Route Traffic Control Centers throughout the country and asked either chief controllers or senior controllers three questions:

1. What are the "bottlenecks," and what is slowing you up in the present air traffic control system?

2. What kinds of equipment, either equipment you have heard about or gadgets that have not yet been designed, would you like to have for use in an air traffic control center?

3. What operations or things done by controllers whom you supervise could be done as well or better by machines or pieces of equipment?

The results of this survey occupy several pages in the Fitts report, and they provide valuable insights into some of the difficulties faced by the men who operate today's equipment.

A study of trucks and buses. A similar kind of study is reported by McFarland and Moseley,[80] who interviewed 100 professional bus drivers and 150 professional truck drivers. Over 300 useful and interesting suggestions were obtained about 13 different classes of design features on modern vehicles. These ranged from comments about visual problems (blind spots, mirrors, and the like) to those about problems of controls (pedals, hand brakes, and so on) and comfort.

Some limitations of opinion data. Despite their potential usefulness, worker or user opinions are often less informative than one might at first suppose. These are several reasons for this:

1. The man who works with a system day-in and day-out may become adapted to the shortcomings of his equipment and so take them for granted. For example, early in World War II, pilots were interviewed for suggestions about improving the aircraft of that time. So few complaints were voiced about the radio sets that one interviewer asked a pilot, "Don't you ever have trouble hearing voice communications on your set?" The reply was, "Sure, but I didn't mention it because I didn't think you could do anything about it."

2. The people who work intimately with a system may become stereotyped in their thinking about it and thus fail to consider bold, radically new ways of accomplishing the same result. This is why an outsider, with a fresh viewpoint, can sometimes make improvements often overlooked by the people too close to the operation.

3. Complaints about machines or methods of work must also be interpreted with caution, because they may only be a symptom of some underlying source of discontent which has little to do with equipment. When people are unhappy—with their foreman, with their pay, with management's handling of grievances—their dissatisfaction may express itself in the form of numerous "gripes" that serve as convenient outlets for accumulated frustration and hostility. For a good discussion of how one must interpret worker complaints refer to Part III of Roethlisberger and Dickson.[95]

4. Collecting valid opinion data is a tricky business and can easily yield deceptive information. The art of asking good questions is one which very few people master. Some ways of asking questions give results which, for all practical purposes, are worthless. Questions like, "Do you enjoy the luxurious ease and personalized service of travel by air?" or "Don't you like the warm tones of this illuminant?" are so heavily loaded that they are of no value whatsoever except perhaps as measures of social conformity under subtle pressure.

Much more difficult to detect and deal with are those slight alterations in wording which change the meaning of questions in important ways. Flanagan,[53] for example, reports that in using the critical-incident technique (discussed in Chapter 3) the last part of one question was: "Tell just how this employee behaved which caused a noticeable decrease in production." This question elicited replies which were almost entirely concerned with matters of personality and attitudes. That part of the question was then changed to read: "Tell just what this employee did which caused a noticeable decrease in production." The second variation produced a much wider range of replies, many of which had to do with the operation of machines. To the person who prepared these questions, "how he behaved" and "what he did" seemed about the same thing. To the foremen who were asked these questions, however, "how he behaved" sounded as if the investigator was primarily interested in studying personality factors. Subtle biases like this can so easily be introduced in the wording of questions that it is always a good idea to try your questions on a typical group of people before you start using them to collect data. Read the questions to each person in the trial group and then ask him to tell you in his own words what he thinks he is supposed to do. This is very often an effective way of uncovering deficiencies in the way you have stated things. If you intend to do any opinion sampling you will also find it well worth your while to study Payne's highly readable book on this subject.[88]

An evaluation of opinion data. For the reasons given above, worker or operator opinions have many important limitations. Nonetheless, the investigator starting out in the business often finds it profitable to begin by collecting such information. If the questions are well framed, and if the answers are taken with the proper degree of skepticism, worker opinions can give the investigator some valuable ideas about where to proceed with more systematic forms of research.

ACTIVITY SAMPLING TECHNIQUES

There are many semiskilled and skilled jobs in life which are complex in the sense that the worker does a number of different things in the course of a day. The housewife, for example, cooks, washes, irons, makes beds, cleans house, sews, supervises children, and works in the garden. A college librarian types, signs out books, looks things up in a card catalog, puts books back on shelves, catalogs new items, consults various reference works, and maintains discipline.

Both of these jobs, moreover, differ from most repetitive industrial tasks in that there is no set pattern in which the various parts of the job are done. The housewife's activities are determined in part by the weather, the day of the week, and variations in the schedules of the other members of the family. The librarian's activities are determined in part by the more-or-less random arrival of people who want to borrow books or ask for assistance. Finally, the times spent in any activity may be highly variable. The librarian usually needs only a minute or two to look up the name of the author of a current book. On the other hand, she may well spend several hours trying to locate some obscure scholarly treatise.

One way of tackling such a complex job is to start by asking, "Exactly how much time does the worker spend in each of his activities?" That is one of the things activity-sampling techniques help us to find out. Activity sampling, as the name suggests, means the systematic observation of an operating system through some sort of sampling procedure. The usual purpose of the sampling is to get an accurate description of what an operator or worker does.

Activity analyses are also used to get descriptions of the activities involved in completing a job or processing a unit of information. In such cases the emphasis is on the job or unit of work, rather than on the operator. This second function is discussed in the section starting on page 37.

SAMPLING OPERATOR ACTIVITIES

The basic technique for this kind of activity sampling is quite simple. The observer has a timing mechanism of some sort. At certain predetermined times the observer records what the worker is doing at that instant. The recording is usually done on a specially prepared report form, two of which will be discussed below (see Figures 3 and 4). When the data are all in, the investigator can then get an estimate of (a) the percentage of the worker's total time spent in various activities, (b) the average length of time spent in each activity, and (c) the sequence in which the worker performs various parts of his job.

With this much as background, let us look at two examples of activity sampling and then see how we actually go about setting one up.

Aerial navigation and radar observing. For an illustration of activity sampling in a military job I have selected a study by Christensen,[32] who recorded the operations performed by aerial navigators and radar operators during fifteen-hour flights. Details of the technique have been discussed in two other articles.[31, 33] Basically, it was this: A timing circuit produced a tone in the observer's earphones every five seconds. Every time the observer heard the tone he recorded what the operator was doing. Observations were recorded in coded form on special sheets. Although the data sheet in Figure 3 is artificial, the sixteen activities

Activity: First Navigator - DC-6B		
Operator: J. Hodgkins	Recorder: S. Wendell	
Time: 0130	Date: 9/7/57	Sampling interval: 5 Sec.
Flight: Reconnaissance squadron A/c No. 456-F		
Remarks: Departed Juneau, Alaska, 0055. Weather cloudy. Enroute to Los Angeles.		

Activity	Tally	Sum	% of Grand Total
Log work	𝖧𝖧 𝖧𝖧 𝖧𝖧 𝖧𝖧 𝖧𝖧 𝖧𝖧 𝖧𝖧 𝖧𝖧	40	16.7 %
Interphone	𝖧𝖧 𝖧𝖧 𝖧𝖧 𝖧𝖧 𝖧𝖧 𝖧𝖧 𝖧𝖧 //	37	15.4 %
Chart work	𝖧𝖧 𝖧𝖧 𝖧𝖧 𝖧𝖧 𝖧𝖧 𝖧𝖧 𝖧𝖧 /	36	15.0 %
Inactive	𝖧𝖧 𝖧𝖧 𝖧𝖧 𝖧𝖧 𝖧𝖧 ///	28	11.7 %
Transition	𝖧𝖧 𝖧𝖧 𝖧𝖧 𝖧𝖧 𝖧𝖧 /	26	10.8 %
Sextant work	𝖧𝖧 𝖧𝖧 𝖧𝖧 ///	18	7.5 %
Eating	𝖧𝖧 𝖧𝖧 ///	13	5.4 %
E-6B computer	𝖧𝖧 ////	9	3.8 %
Map reading	𝖧𝖧 ////	9	3.8 %
Astrocompass	𝖧𝖧	5	2.1 %
Auxiliary radar	////	4	1.7 %
Radio	///	3	1.2 %
Altimeter	//	2	0.8 %
Drift reading	/	1	0.4 %
Other activity	𝖧𝖧 ////	9	3.8 %
Grand total		240	100.1 %

FIG. 3. One type of data sheet used in activity analyses. This data sheet is fictitious, but the activities in the left-hand column are those used in a study by Christensen.[32] The percentages in the right-hand column agree almost exactly with those found by Christensen for the activities of the first navigator.

listed in the left-hand column are those actually used for the first navigator. Similar data sheets were prepared for observing the second navigator and radar operator.

The percentages shown in Figure 3 are almost exactly what Christensen actually found for the first navigator's job, although, of course, his observations were much more extensive than those shown here. One of the interesting, and unexpected, findings of this study was that the navigator spent a very large proportion of his time in paper-work—keeping the log and working on charts. Together these two tasks took up nearly a third of the navigator's time. This fact alone suggested that it might be profitable to examine all this paper work critically with an eye to simplifying this part of the job—by eliminating needless repetitive recording of data, by eliminating the recording of data that are never used, and by providing more efficient plotting tools and charts. Other kinds of improvements were suggested by other aspects of the data. The final outcome of this study was that Christensen was able to reduce the number of men in the aircraft by two without over-burdening any of the remaining crew members. These gains were entirely the result of better workplace arrangements, more efficient procedures, and new equipment aids.

The activities of bus driving. The example just discussed is concerned with rather large units of activity—things like reading maps, doing log work, and computing the drift of the aircraft. A study by McFarland and Moseley (pp. 228 ff.) shows that activity sampling can be applied to very much smaller units of activity.[80] In their study, observations were made between Boston and New York on scheduled interstate bus trips and were recorded during 7 of the 9½ hours required for each trip. Data were collected on the eye, head, body, hand, and foot movements used in turning, decelerating, accelerating, shifting, stopping suddenly, and braking. Table 1 shows only a small part of the data collected in this study. You can see, incidentally, that the right hand engages in many more activities than does the left hand.

Planning the activity analysis. The preceding discussion suggested some of the things—the timing mechanism and data sheet—that need to be considered in planning an activity analysis. Here is a more systematic listing of the various steps in setting up such a study:

1. ESTABLISH RAPPORT with and secure the co-operation of the subjects who will be observed. The reasons for this should be obvious. Before you start following people around with a stop watch and pad, you have to explain what you are doing and why. If for no other reason,

this is a courteous gesture which will allay feelings of hostility and suspicion. In some industries, in fact, it may be absolutely essential to secure the permission of the union steward before you make any observations of any sort. Another reason for this personal contact is that you want the people being observed to behave as naturally as possible.

2. STUDY THE ENTIRE JOB to become familiar with all the steps and all the activities. Complete familiarity with the job is essential so that

TABLE 1. *Activity analysis of the right and left hands in bus driving. This table shows only a part of the data collected by McFarland and Moseley*[80]

Activity	Right hand		Left hand	
	Frequency	Per cent	Frequency	Per cent
Wheel	1200	80.9	1465	98.8
Shift	161	10.8	0	0.0
Rest	97	6.5	17	1.1
Hand brake	4	0.3	0	0.0
Instruments	6	0.4	0	0.0
Fares	3	0.2	0	0.0
Signals	0	0.0	0	0.0
Horn	13	0.9	0	0.0
Waving	0	0.0	2	0.1
Total	1484	100.0	1484	100.0

you can tell immediately what the operator is doing when he looks through that eyepiece, twiddles that dial, or punches that series of buttons. This is also an essential prerequisite to setting up categories of observation, the topic which follows immediately below.

3. DECIDE ON AND SET UP CATEGORIES OF OBSERVATION. Some typical categories of observation are illustrated in Table 1 and Figure 3. There are a number of factors that should influence your decision:

(a) How coarse or how fine should the units of activity be? To a large extent, the answer to this question is determined by the purpose of the analysis and what the investigator hopes to find out. The two examples given above differ markedly in this respect. In the study of the aerial navigator, the units of activity were gross, doing log work, eating, and using the interphone. In the study of bus drivers, the units of activity were very much smaller, being entirely concerned with what the drivers' hands were doing. You can probably see at once why it would not have been of very much value to have done such a fine-grained analysis of the navigator's job.

(b) The categories should exhaust all the activities the operator engages in. Although there should be a special category for "other activities," you should try to end up with as few observations as possible in this class. Unless you include all of the things the operator does, you cannot compute meaningful percentages.

(c) Categories should be of observable behavior (refer to the discussion on pages 18–21). Since this is an especially important point, let me elaborate on it in this way. The categories should be of activities which everyone can agree on. They should not include such things as "thinking," "daydreaming," "planning," or "doing mental problems." The trouble with designations like these is that they force the investigator to make an inference about the operator's behavior. When an operator is sitting with a vacant expression on his face, he might be doing any of these things. You have no way of knowing, and there is no point in thinking that you do. In short, try to pick activities which are clear-cut and can be interpreted with little or no uncertainty.

(d) Be sure to include transitions from one task to another. A transition is the time an operator takes to put away old work, to go from one work place to another, or to prepare himself for a new task.

(e) Define the limits of each activity. For example, in an activity analysis of a secretary's job, is she "filing" only when she is literally inserting pieces of paper into a file? Or does "filing" also include "unlocking the files" and "opening the file drawers"? You could, if you wanted, define "unlocking the files" as a "transition" activity. It often does not make very much difference how narrowly or broadly one defines many of these activities. The important thing is to have a clear-cut definition of each activity so that you will know how to classify everything you see with no ambiguity.

(f) The last point is that you should not set up too many categories for observation. In general, if you have more than about twenty-five different activities to observe, you may find yourself getting confused, forgetting the symbols, or getting lost on the data sheet. Do not make the job of observation too hard.

4. SET UP THE DATA SHEET. Figure 3 illustrates one way of setting up a data sheet; Figure 4 another. The former is simpler than the latter, but it does not give the investigator as much information. The data sheet in Figure 3 lists the various activities the operator engages in, and the observer is required merely to make tallies in the appropriate places. For example, if the operator is using the radio at the first time signal, the recorder makes a tally in the twelfth row. If the operator is using the interphone at the second time signal, the observer makes a tally in the second row. And so on. When the observation period is over the investigator has a good picture of the percentage of the total time the operator spent in each activity. But that is all this kind of analysis tells you.

The data sheet in Figure 4 is a little more difficult to use and requires more time to analyze, but it also yields more kinds of data. Note that each of the sixteen activities has a special code. The observer records in the squares below exactly what the operator is doing at every five-second interval. Appropriate summaries of such data provide estimates of (a) the total time spent in each activity, (b) the sequence in which the operator performed various activities, and (c) the percentage of the total time spent in each activity. For example, you can see that when the observation period started, the operator was using the radio. He then used the phone for at least fifteen seconds, spent a little time getting ready to take a drift reading, and then spent over a minute and a half actually taking a drift reading. This was followed by a short transition period, after which the operator made some notes in his log, and so on. Each line of this data sheet represents two minutes of total observation. Altogether four minutes of observation are shown there.

5. DECIDE ON A SAMPLING DURATION. By this I mean the total duration through which observations will be made. The decision may be to do either of two different things: (a) you may sample an entire unit of activity (for example, one flight, one trip with a long-haul truck) if the activity comes in such convenient units; or (b) you may sample a convenient unit of time (for example, two hours, one day, one month, or ten one-hour periods distributed throughout a week).

If the sampling duration is made on the basis of time—that is, (b) above—the time units should be selected to sample fairly all of the different kinds of activities which can occur. Certain kinds of tasks are often typically done at certain times of the day. The housewife, for example, may do most of her baking in the afternoon, and if the time samples were selected from only the morning hours, the data would

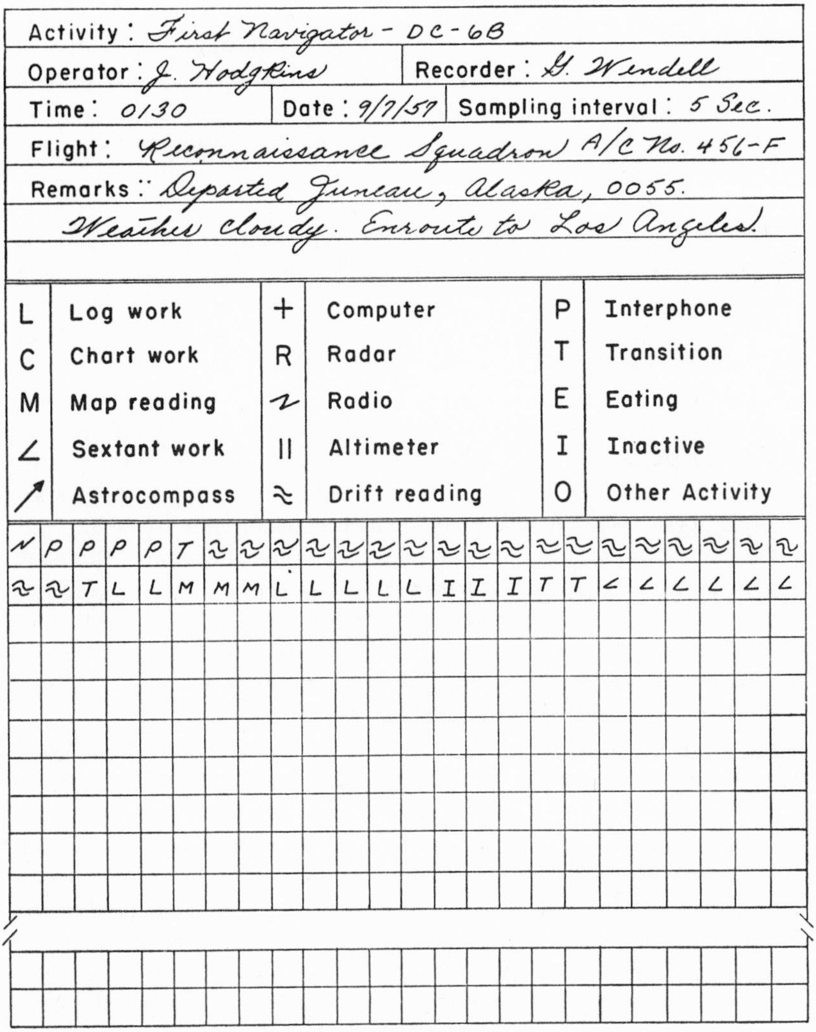

Activity: *First Navigator - DC-6B*

Operator: *J. Hodgkins*   Recorder: *G. Wendell*

Time: *0130*   Date: *9/7/57*   Sampling interval: *5 Sec.*

Flight: *Reconnaissance Squadron A/C No. 456-F*

Remarks: *Departed Juneau, Alaska, 0055.*
Weather cloudy. Enroute to Los Angeles.

L	Log work	+	Computer
C	Chart work	R	Radar
M	Map reading	∿	Radio
∠	Sextant work	‖	Altimeter
↗	Astrocompass	≋	Drift reading

P	Interphone			
T	Transition			
E	Eating			
I	Inactive			
O	Other Activity			

↗	P	P	P	P	T	∿	∿	∿	∿	∿	∿	∿	∿	∿	∿	∿	∿	∿	∿	∿	∿
∿	∿	T	L	L	M	M	M	L	L	L	L	L	I	I	I	T	T	∠	∠	∠	∠

FIG. 4. Another type of data sheet used in activity analyses. This data sheet is fictitious, but the activities are those used in a study by Christensen.[32]

not be representative. In Christensen's study, to take another example, there was a gradual but progressive decline in the number of drift readings the navigator made throughout a long flight. This appeared to be primarily a matter of fatigue. For our purposes here the important point to notice is that the navigator distributed his time somewhat differently at the beginning of a ten-hour flight than at the end. If Christensen had studied only the first couple of hours of flight he would have finished with a biased picture of how the navigator spends his time.

6. DECIDE ON A SAMPLING INTERVAL. The sampling interval is the time between successive observations. In the Christensen study of aerial navigators, for example, the sampling interval was five seconds. Actually, the first choice is between a random (that is, variable) and fixed sampling interval. Random sampling intervals can be used when the investigator is interested only in the distribution of operator activities. It will not allow the investigator to compute the average lengths of time spent in various activities, or the sequences of activities the operator engages in. In addition, random sampling intervals are generally best suited to operations which continue day after day and which can be observed over long periods of time (examples: the work of the housewife, librarian, secretary, stock clerk). Fixed sampling intervals are best suited to jobs which are variable in length and which have a clearcut beginning and end (examples: the work of the navigator or bus driver). Examples of jobs done with random sampling intervals are given in Heiland and Richardson.[64]

A lower limit to fixed interval values is about two seconds, a figure which is set primarily by the ability of an observer to record various activities. If the interval is shorter than two seconds, the observer is likely to get lost, fall behind, and become inaccurate in his observations. In any event, the investigator will rarely want to make observations as rapidly as this because most activity sampling studies are concerned with rather gross units of activity which come in larger time blocks. If you are concerned with very small units of activity, you should consider using memomotion techniques, discussed in the next section.

If you want to get a complete description of the sequences of activities engaged in by an operator (as illustrated in Figure 4), then the sampling interval should not be longer than the shortest unit of activity. If the sampling interval is longer than certain kinds of activity, they will be missed completely whenever they fall between two signals.

If you only want an estimate of the relative proportion of time spent
by an operator in various kinds of activities (as illustrated in Figure 3),
then the sampling interval may be longer than the shortest activity.
Although small segments of activity will occasionally be missed with
such a scheme, they should show up in their proper proportions if the
overall sampling duration is long enough.

The final consideration is that for a worthwhile study the investigator should end up with a sufficiently large number of observations to
get valid and reliable data. As a very rough guide, perhaps something
on the order of 1,000 observations should be the minimum number
collected. This in turn suggests an interaction between sampling duration and sampling interval. The product of sampling duration and
sampling rate (reciprocal of sampling interval) should be at least 1,000.
For example, if the sampling interval is five seconds, the sampling rate
is 12 per minute and the sampling duration should be a minimum of 83
minutes. If the sampling interval is 1 minute, the sampling rate is 1
per minute, and the sampling duration should be at least 1,000 minutes
long.

7. SELECT A REPRESENTATIVE SAMPLE OF SUBJECTS. People differ in
the way they carry out the same task, and if the results of an activity
analysis are to be useful the investigator should study several operators. Moreover, he must assure himself that the operators he has
selected for study are representative, or typical, of the operators who
normally carry out these tasks. The problem of individual differences
among people and human sampling is more thoroughly discussed in
Chapter 6.

8. SELECT A REPRESENTATIVE SAMPLE OF JOBS AND TASKS. Although
most people readily recognize that there are large individual differences
between operators in the way they do things, the problem of *job
sampling* is not often recognized. The housewife's task depends greatly
on the number of children she has, the size of house she has, and the
layout of the house; the librarian in a small library has an entirely
different kind of job from the one in a large library; one aircraft differs
from another; and secretarial duties in one office differ from those in
the next.

If the activity analysis will be used only for making general statements about one particular house, one particular library, or one particular aircraft, then there is no serious problem of job sampling. But if
the results of the activity analysis will be used to make statements
about housewife activities in general, librarians in general, or flying

aircraft of this general type, then the investigator must be sure to take measurements on a representative sample of equipment, jobs, or installations. This means that he has a general problem of job sampling as well as one of operator sampling.

To provide a concrete illustration we return to Christensen's study of the aerial navigator. Navigating in polar regions is quite a different job from navigating in lower latitudes. In the arctic there are few prominent terrain features, no man-made guideposts, and erratic behavior from magnetic compasses. As a result, the pattern of the navigator's activities is very much different in these two locations. The investigator needs to take such factors into account if he wants to get a fair picture of the operator's job.

MEMOMOTION STUDY

Memomotion study is a special form of activity sampling in which (a) the sampling interval is usually shorter than two seconds, and (b) a motion-picture camera replaces unaided visual observation. It is one of the standard techniques developed by industrial engineers for studying various kinds of jobs.

In memomotion studies, motion pictures are taken at unusually slow speeds, one frame per second being a common value. When the film is developed each frame is studied, and the operator's activities are then catalogued in exactly the same way as in Figures 3 or 4. All of the other considerations discussed on pages 29–36 apply equally to memomotion study.

Some advantages and disadvantages of memomotion study. Compared with a human observer for activity sampling, memomotion study has both advantages and disadvantages. Its principal advantage is that it provides a permanent record which allows the investigator to examine and re-examine the operator's actions. A second advantage is that it enables the investigator to take observations more frequently than would be possible using a human observer (see the discussion on sampling intervals on page 34).

There are, however, some advantages in using a human observer for activity sampling. First and foremost, a human observer requires little in the way of equipment and is versatile. For example, Christensen could make observations during night flights under conditions which would be difficult to record photographically. In addition, the observer could take off in any one of several aircraft with very little preparation. If memomotion techniques had been used, preparations would have to be made long in advance of take-off. Finally, the human observer is more flexible than a motion-picture camera—especially within the crowded confines of modern military aircraft. The observer can follow the operator around, and the operator does not need to confine his activities to the field of view of the camera.

ACTIVITY SAMPLING ON A UNIT BASIS

The two preceding sections discussed activity sampling from the standpoint of all the things particular workers do. In this section we shall take a different point of view and look instead at all the things that get done to a particular unit of work or product. To emphasize this distinction, let us consider the following two ways we could study central telephone activities.

If our orientation were the activities of the telephone operator, we would fashion our activity sampling study around the various things the operator does—she receives incoming calls, looks up telephone numbers, dials numbers, waits for subscribers to answer the phone, and makes out toll tickets, for example. This would be an activity sampling study of the sort discussed in the two previous sections. To take quite a different point of view, however, we could focus our attention on the history of typical long-distance telephone calls. Now we would trace telephone calls through the various steps required to connect a subscriber with the person he is calling. In this case we would probably measure such items as the length of time it took the subscriber to reach the operator, the time it took the operator to look up the number, the time it took her to dial the distant subscriber, and the time it took for the distant subscriber to answer his phone.

The composition of newspaper advertisements. A good illustration of this form of activity sampling comes from some work done at Johns Hopkins for the Institute of Newspaper Operations. Advertisements provide a major part of the income for most newspapers, but in these days of rising costs the profit margin from this source of revenue is narrowing. Thus, newspapers are generally concerned with increasing the efficiency, and decreasing the costs, of composing and printing advertisements. As a first step in this direction the Hopkins investigators measured how long it took to prepare and to print typical local display advertisements (the large advertisements with pictures that department stores, grocery stores, and automobile dealers typically insert into newspapers). The reason for the study was that most newspapers had no idea how much time was spent in the preparation of such advertisements; and time, of course, is one of the most important elements contributing to cost.

There are a large number of separate steps involved in making-up and composing local display ads. Although it is not essential that we understand all of the operations, we need to get a rough idea of some of the steps involved. Typically, the customer comes in with only a rough pencilled sketch of some illustrations, some text describing the merchandise and some prices. These notes and copy are then usually sent to a mark-up man who does a systematic job of planning and laying out the illustrations, headings, and text. From here it goes to the linotype operator who sets in type whatever text there is in the ad. Next it goes to the Ludlow operator who prepares special headlines and captions. From there it goes to the assembly desk, where headings, text, and illustrations are assembled. Proofs are then made of the complete ad and errors corrected. The ad next goes to the customer for his corrections, then back to the newspaper so that the customer's changes can be incorporated into the ad.

Without any further elaboration you can see that a lot of different things happen to an ad before it appears on page 3 of your morning paper. Perhaps you can also visualize the procedure used in this study. An advertisement was timed through each of these operations so that the investigators established its complete time history. They knew how long it had been in mark-up, Lino, Ludlow, assembly, and revision. By studying a large sample of such ads—of various sizes, and complexities—they were then able to see that the newspapers were not charging enough for certain ads. The time spent in composing and printing some ads (and hence the cost of printing them) was actually greater than the revenues derived from them. Another finding was that one operation (office proofreading before proofs were sent to the customer for correction) could be eliminated with a substantial saving in time.

Telephone service observing. It is possible, of course, to collect data on one part of a larger job and to take error measurements rather than time measurements. Such a variation of activity sampling is the service-observing that goes on continually throughout the telephone system.

In the telephone centers of most large cities there is a special monitoring station from which observers can listen in on the telephone calls handled by any of several boards, each of which is manned by a long-distance operator. Although the observer is not allowed (by law) to listen in on conversations, she can hear the subscriber placing the call with the operator, can hear the operator's reply, and can then see on a display board what the operator dials. From such service-observing it has been possible to discover important sources of human troubles. For example, in one period of 1,120 calls observed in Washington, D. C., 12.5 per cent of the calls were in error due to the operator. One of the largest categories of error was due to operators hitting the wrong key on their keyboards.

In this example, notice that only a part of the telephone operator's job is being sampled. Actually the operator does more than just dial numbers. She also makes out toll tickets, for example, from which time and charges are computed. But in this case, only the dialing part of the job is studied. The other thing to notice in this example is that the investigator does not find out what proportion of the operator's time is spent in dialing numbers. Rather he finds out the number and kinds of mistakes operators make in completing long-distance calls. To adapt this technique to Christensen's study, Christensen might, for example, have double-checked all the navigator's drift readings to get some indication of the number of errors that are made in practice. This would have given him a different sort of data—but useful data nonetheless.

PROCESS ANALYSIS

Process analysis refers to a group of techniques for recording compactly the various steps involved in a process. In a factory, for example, the process begins when raw material enters the plant. Inside the factory the material may be transported, stored, inspected, machined, assembled, and inspected again. The process is usually considered complete when the finished items are stored for shipment. The analysis can, of course, record a process through only one or a few departments. In an office, process analysis might be used to study the flow of time cards, materials requisitions, purchase orders, or even correspondence. In general, process analysis techniques are most useful in problems involving the layout or arrangement of work and equipments. Four techniques are considered below. They are process charts, flow diagrams, multiple-process charts, and link analysis.

Process charts are useful for describing the various steps involved in routine or standardized operations. They may be used to describe (a) the various stages in the manufacture or processing of a particular product or (b) the various activities an operator goes through in doing his job.

Basic symbols. Since there are some variations among the kinds of symbols used by different time-and-motion engineers, those illustrated in Figure 5 have been selected to combine the best features of several systems. Note that there are six different symbols illustrated there. Their definitions are as follows:

1. OPERATION. A large circle indicates that something is done to a product at essentially one location, or, if the primary purpose of the process chart is to analyze the activities of an operator or worker, it refers to an operator doing something at one fixed place. For most jobs this symbol identifies the main activities in the entire process. Things like transportation, movement, storage, and inspections are generally considered to be auxiliary to these steps in the process.

2. TRANSPORTATION. Small circles indicate transportation or the movement of objecɩs being studied from one position or location to another. If the symbol is used to describe an operator's activities, it would indicate a movement of the operator from one place to another. Sometimes it may be desirable to show the means by which a move was accomplished. This is often done by placing a letter inside the circle. For example: M = man; H = handtruck; K = power truck; C = conveyor; E = elevator; L = mail. When material is stored beside or within two or three feet of a bench or machine on which an operation is being performed, the movements used in getting the material before the operation and in disposing of the piece to another box are generally considered to be part of the operation.

3. TEMPORARY STORAGE. The double triangle indicates a temporary storage or delay. In the case of a product it refers to the product being temporarily held at one location without an operation being performed on it. If an individual rather than a product is being charted,

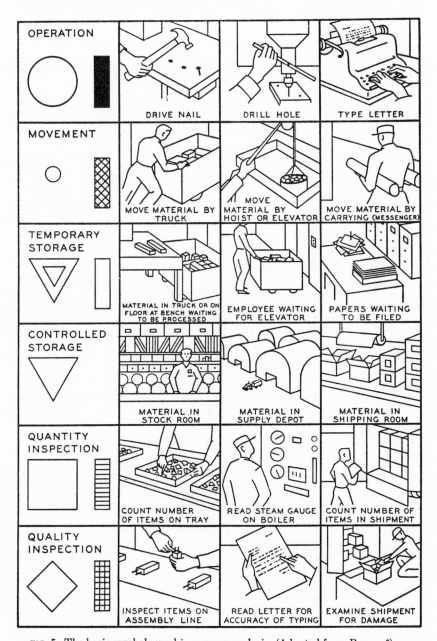

FIG. 5. The basic symbols used in process analysis. (Adapted from Barnes.[9])

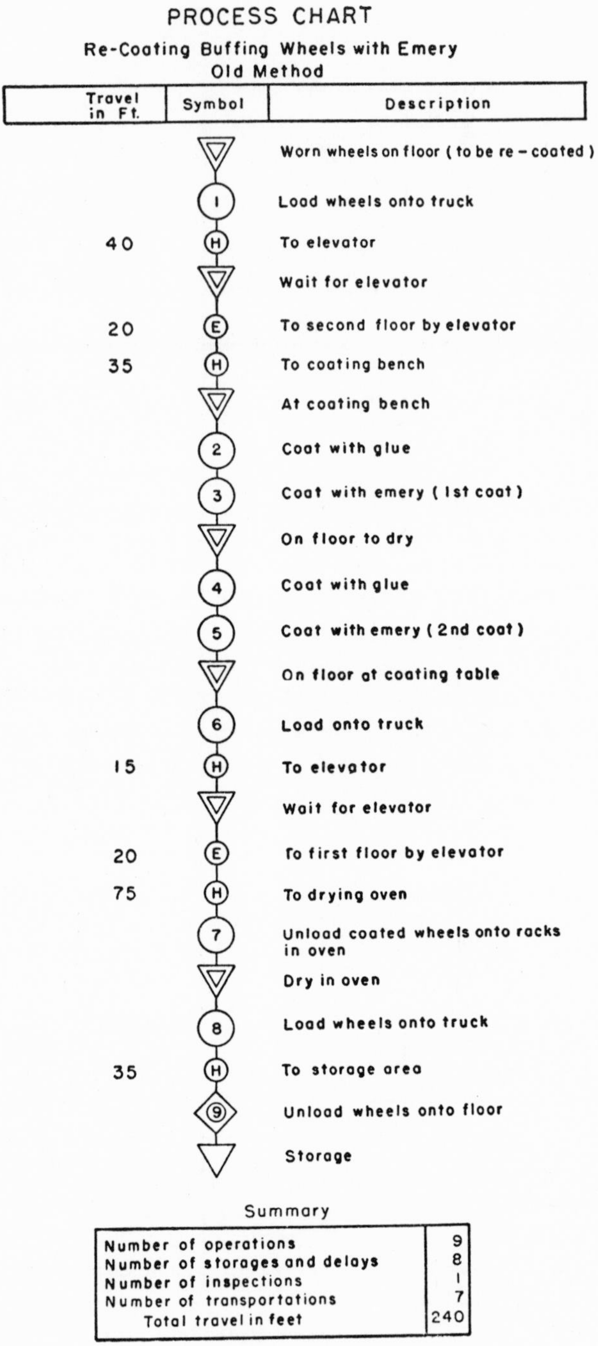

PROCESS CHART
Re-Coating Buffing Wheels with Emery
Old Method

Travel in Ft.	Symbol	Description
	▽	Worn wheels on floor (to be re – coated)
	①	Load wheels onto truck
40	Ⓗ	To elevator
	▽	Wait for elevator
20	Ⓔ	To second floor by elevator
35	Ⓗ	To coating bench
	▽	At coating bench
	②	Coat with glue
	③	Coat with emery (1st coat)
	▽	On floor to dry
	④	Coat with glue
	⑤	Coat with emery (2nd coat)
	▽	On floor at coating table
	⑥	Load onto truck
15	Ⓗ	To elevator
	▽	Wait for elevator
20	Ⓔ	To first floor by elevator
75	Ⓗ	To drying oven
	⑦	Unload coated wheels onto racks in oven
	▽	Dry in oven
	⑧	Load wheels onto truck
35	Ⓗ	To storage area
	◈⑨	Unload wheels onto floor
	▽	Storage

FIG. 6. Process chart of a buffing wheel operation. (After Barnes.[9])

Summary

Number of operations	9
Number of storages and delays	8
Number of inspections	1
Number of transportations	7
Total travel in feet	240

this symbol would be used to indicate idleness on the part of the operator.

4. CONTROLLED STORAGE. A large open triangle indicates the storage of the product in a stock room, warehouse, or shipping room. Items in controlled storage ordinarily cannot be moved without an order, shipping ticket, or some other kind of record.

5. QUANTITY INSPECTION. A square indicates an inspection for quantity. This may consist of measuring, counting, or weighing. It is essentially any operation in which things are counted or in which quantitative information is obtained.

6. QUALITY INSPECTION. A quality inspection usually consists of testing a part to see whether it will pass a predetermined standard. It also may take the form of grading or examining items for damage.

Process chart of a buffing-wheel operation. In large factories where there are numerous heavy polishing and buffing operations it is customary to recoat buffing wheels in the plant in order to have a supply of fresh wheels always available. The method involves coating the circumference of the worn wheel with glue and then rolling the wheel by hand through a shallow trough filled with emery dust. After the glue has dried a second coat of glue and emery dust is applied. The wheels are then carted to a drying oven where they are hung on racks in the oven until the glue is thoroughly dry. Figure 6 shows a process chart of this procedure as it was carried out in one factory.

FLOW DIAGRAMS

The process chart in Figure 6 shows in symbolic form the various steps involved in recoating buffing wheels but does not show *where* the various steps of this procedure were carried out. A flow diagram is a graph of a process chart showing the locations of all of the operations. Such a flow diagram for the process in Figure 6 is shown in Figure 7. There you can see that the original process for recoating buffing wheels was carried out on two separate floors with the various parts of the activity widely dispersed. A revised and greatly improved procedure for accomplishing the same result was designed by placing the coating operation in the corner formed by the storage area and oven on the first floor. This reduced the time and effort required to carry the wheels to the second floor and back by way of an elevator. It also greatly shortened the distances required for transporting the wheels from the storage area to the oven and back to the storage area again.

Flow diagram of a cockpit checkoff procedure. Flow diagrams need not be as detailed as that shown in Figure 7. Sometimes a much simpler form of diagram can suggest improvements in a given procedure. As an example, let us look at a study carried out on a multi-engine aircraft.

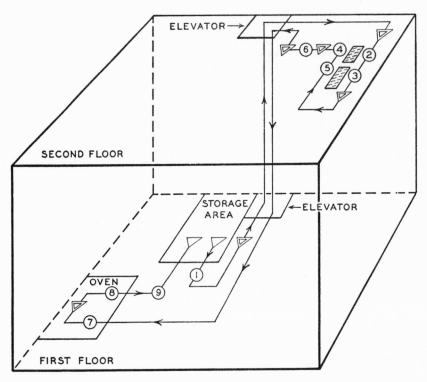

FIG. 7. Flow diagram of the process chart in Figure 6. (After Barnes.[9])

The casual observer who steps into the cockpit of a multi-engine aircraft is confronted with a maze of instruments and controls. Closer study of the entire operation of flying an aircraft further impresses the observer of the complexities involved in the task. As one way of studying the pilot's and co-pilot's task in flying an aircraft, Channell and his co-workers divided the cockpit into seven general work areas as follows:[18]

Area 1—Control column, rudder and brake pedals
Area 2—Control pedestal (upper half)

Area 3—Control pedestal (lower half)
Area 4—Main instrument panel (lower portion)
Area 5—Main instrument panel (upper portion)
Area 6—Pilot's upper instrument and switch panels
Area 7—Side control panel

It is not really necessary to describe in any greater detail what these areas encompass, for we are concerned here only with points of procedure, which can be understood without identifying the precise instruments involved.

Before taking off, the pilot and co-pilot go through a cockpit check-off. That is, they systematically check instruments, controls, and functions of the aircraft to be sure that everything is in proper working order. The sequence of operations which was prescribed for check-off in this aircraft involved paths of movement as illustrated in the upper part of Figure 8. For example, the pilot first had to check the battery switch in Area 6, then the instrument switch in Area 6, then the wing tanks in Area 2, the carburetor air in Area 3, the cross feeds in Area 2, and so on.

A simple flow diagram of the paths of movement required by the pilot in executing the entire check-off reveals a very complicated procedure involving considerable and needless movement back and forth between areas. The lower part of Figure 8 shows a simplified sequence of movements developed by rearranging the order of check-off of the various instruments. In this simplified scheme the controls in any one area are completely checked before moving on to the next area, thus providing for a more smoothly flowing path of movement. In addition, the order of checking controls within each area can be standardized to reduce the likelihood of omissions during the check-off procedure.

For purposes of our discussion here, however, it is important to notice that this is a flow diagram of essentially the same sort as was illustrated in Figure 7 except that only paths of movement are shown between the various work areas. Even so, such a simplified flow diagram was very helpful in redesigning and increasing the efficiency of this operation.

MULTIPLE-PROCESS CHARTS

Multiple-process analysis is another variation of general process analysis. Two things distinguish it from the ordinary techniques dis-

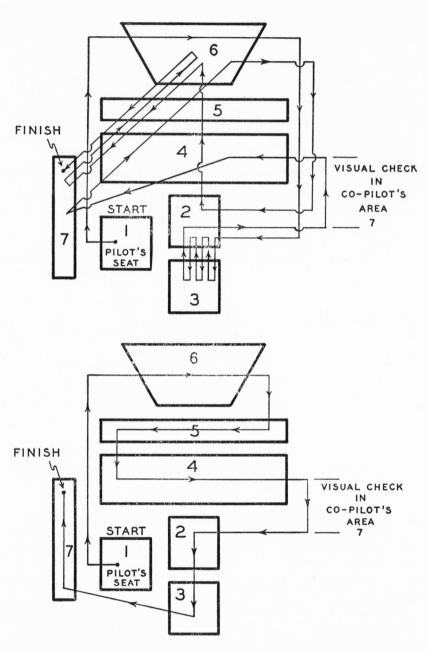

FIG. 8. The upper part of the figure is a flow diagram of the check-off procedure used by a pilot in preparing a plane for take-off. The lower part of the figure shows a flow diagram of the improved procedure. (After Channell.[18])

THREE **MAN** AND ONE MACHINE **PROCESS CHART**

METHOD -1; CUTTING STUDS ON DE-WALT -M 127

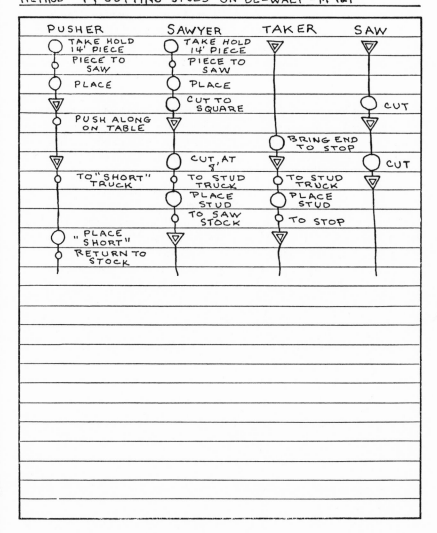

PUSHER	SAWYER	TAKER	SAW
○ TAKE HOLD 14' PIECE	○ TAKE HOLD 14' PIECE	▽	▽
○ PIECE TO SAW	○ PIECE TO SAW		
○ PLACE	○ PLACE		
▽	○ CUT TO SQUARE		○ CUT
○ PUSH ALONG ON TABLE	▽		▽
		○ BRING END TO STOP	
▽	○ CUT AT 8'	▽	○ CUT
○ TO "SHORT" TRUCK	○ TO STUD TRUCK	○ TO STUD TRUCK	▽
	○ PLACE STUD	○ PLACE STUD	
	○ TO SAW STOCK	○ TO STOP	
○ PLACE " SHORT"	▽	▽	
○ RETURN TO STOCK			

FIG. 9. Multiple-process chart showing the operation of a three-man crew cutting 8-foot studs. (After Mundel.[84])

cussed above: (a) it has time values added so that you can see how long each step occupied; and (b) it is ordinarily concerned with multiple processes, several things going on at the same time. The value of the multiple-process chart is that it shows what elements of the different operations occurred at what times. This, in turn, can be helpful in suggesting how times can be rescheduled for each part of a complex operation to produce a more efficient utilization of men and machines.

Multiple-process analysis of a sawing operation. Figure 9 is a process chart showing the operation of a three-man crew cutting 8-foot studs from 14-foot lengths of 2 × 4 lumber. On this chart one column has been used for each man and one column for the saw. The various symbols are keyed roughly in time—all of the operations on any line occur at the same time. However, the distances in any column are not proportional to time.

The data in Figure 9 are replotted in Figure 10 in a man-machine process chart. Here the lengths of the various segments in each column are proportional to time. The different markings have the meanings defined in Figure 5. The easiest way to obtain these values is from motion picture films because such records enable the investigator to obtain all four times simultaneously. Alternative but less satisfactory methods of collecting the data are to use multiple observers, one for each man and machine being observed, or to time the operators and machines in successive cycles of operation.

Other variations. There are many variations of this basic procedure, the chief difference between them being merely what one plots in the various columns. For example, one may prepare charts for one man and one machine, for one man and several machines, or for the right and left hands of a single operator. In every case the essential technique is the same as has been described above. Whatever the method used for collecting the basic data, or whatever the process being studied, many cycles of operation should be observed and average times computed. Refer again to the discussion on pages 29–36, because many of the points discussed there apply to the collection of data for this purpose as well.

Job sampling again. In discussing these techniques the authors of most textbooks seem to ignore the problem of job sampling. The reason for this is undoubtedly that time-and-motion techniques were originally devised for jobs which were highly repetitive and standardized. As a result, the variability from job to job was probably comparatively small. Even so, it is safe to assume that an operator will never drill

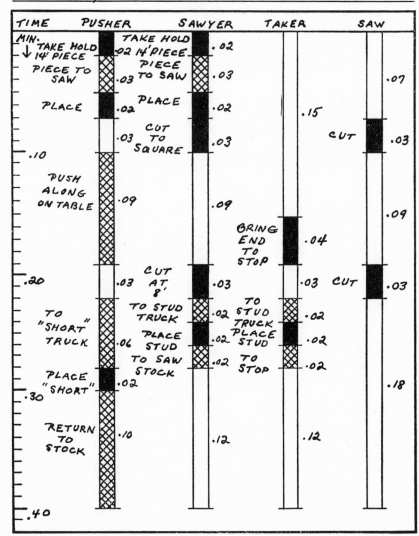

FIG. 10. Man-machine process chart of the operation illustrated in Figure 9. (After Mundel.[84])

PROCESS CHART

Date: JUNE 25, 1946   Type of Plane: R5D   Flight Phase: LANDING, INCLUDING PART OF FINAL APPROACH

Place: GUANTANAMO   Flight conditions: DAYLIGHT, ROUTINE

PILOT						CO-PILOT		
				TIME (MINUTES)				
SEQUENCE OF OPERATIONS	HAND USED	WORK AREA	EACH OPER-ATION	CUMU-LATIVE	EACH OPER-ATION	WORK AREA	HAND USED	SEQUENCE OF OPERATIONS
HANDS ON THROTTLE	R	2	.037	.05	.062	1	L,R	RESTS ARM ON ARM REST
RETURNS HAND TO YOKE	R	1	.116	.10				
				.15	.121	3	L	REACHES FOR LANDING GEAR
				.20				
				.25				
REACHES FOR THROTTLE AND OPERATES THROTTLE	R	2	.440	.30	.164	3	L	PUSHES GEAR HANDLE DOWN
				.35	.017	1	L	MOVES HAND TO FLAP LEVER
				.40	.102	3	L	OPERATES FLAP HANDLE
				.45				
				.50	.047	1	L	RETURNS TO ARM REST
					.020	3	L	ADJUSTS FLAP HANDLE
				.55				
ADJUSTS MIXTURE CONTROL	R	3	.020		.051	1	L	RETURNS ARM TO REST
				.60				
RETURNS HAND TO THROTTLE	R	2	.125	.65				
ADJUSTS MIXTURE CONTROL	R	3	.003	.70	.252	3	L	ADJUSTS FLAP
RETURNS HAND TO THROTTLE	R	2	.040	.75				
RETURNS HAND TO YOKE	R	1	.080	.80				
				.85				
ADJUSTS THROTTLE	R	2	.080	.90	.010	1	L	RETURNS HAND TO ARM REST
				.95				
RETURNS HAND TO CONTROL COLUMN	R	1	.075		.075	2	L	OPERATES NWS SET OF THROTTLES
ON DECK			.003	1.00	.019	1	L	RETURNS HAND TO ARM REST
REACHES FOR THROTTLE	R	2			.021	3	L	REACHES FOR FLAP HANDLE
				1.05	.052	3	L	RAISES FLAP HANDLE
				1.10	.028	1	L	RETURNS HAND TO ARM REST
					.012	3	L	ADJUSTS FLAP HANDLE
				1.15		1	L	RETURNS HAND TO ARM REST

FIG. 11. Process chart showing the activities of a pilot and co-pilot in landing an R5D aircraft. (After Channell.[18])

two holes, assemble two relays, or solder two connections in precisely the same way. This means that if an investigator wants to get a reasonable description of a task, he needs to sample it many times. Unless he does so, in fact, he may find that he has spent much time worrying about a process chart which is not at all typical of the operation.

As an illustration of the amount of variability that can occur in tasks which are not as routine as industrial jobs, look at the two process charts in Figures 11 and 12. These show the activities of the pilot and co-pilot in landing an R5D aircraft. The aircraft was the same in both cases, the weather and the flight conditions were approximately the same (daylight, routine), and the records were taken within a few weeks of each other. Yet even a cursory inspection of these two charts shows such wide divergences that it is difficult to believe they refer to the same operation.

LINK ANALYSIS

The flow-process chart and flow diagram were developed primarily for analyzing work situations which are repetitive and standardized. For example, recoating buffing wheels is a standardized procedure, since operators always go through the same routine and sequence of operations. Similarly, the preflight check-off procedure used by the pilot and co-pilot is carried out in exactly the same order on every flight.

But there are many jobs in the world that are not so routine. Even driving a car is not a simple 1–2–3 proposition. Sometimes we push the clutch before the brake, and sometimes the brake first. Sometimes we step on the accelerator before we turn the wheel, and sometimes we slow down first. A man in an aircraft control tower does many different things, and he does not always do them in the same order. The same is true of the radar operator, the pilot of an aircraft, and operators in many other tasks which are not simply a matter of assembly line production. Link analysis is a technique for handling situations of this sort.

A link analysis is a kind of flow diagram in which the linkages between various components are expressed in statistical terms—relative frequencies between components. It is useful primarily in problems of layout and arrangement of men and machines. It is not concerned with how long a man spends at a piece of equipment or how he operates it.

116495

PROCESS CHART

Date: JUNE 3, 1946 Type of Plane: R5D Flight Phase: LANDING, INCLUDING PART OF FINAL APPROACH

Place: WASHINGTON, D.C. Flight conditions: DAYLIGHT, ROUTINE

PILOT					CO-PILOT			
				TIME (MINUTES)				
SEQUENCE OF OPERATIONS	HAND USED	WORK AREA	EACH OPER-ATION	CUMU-LATIVE	EACH OPER-ATION	WORK AREA	HAND USED	SEQUENCE OF OPERATIONS
				.05	.110			IDLE
REACHES FOR THROTTLE TO REDUCE POWER OPERATES THROTTLE	R	2	334	.10				
				.15	.130	3	L	LOWERS FLAP HANDLE
				.20				
				.25				
				.30	.100	1	L	RETURNS HAND TO ARM REST
RETURNS TO CONTROL COLUMN	R	1	.005					
ADJUSTS ELEVATOR TRIM	R	2	.020	.35				
RETURNS TO CONTROL COLUMN	R	1	.003					
				.40				
REACHES FOR THROTTLE OPERATES THROTTLE	R	2	.132		.124			IDLE
				.45				
					.031	2	L	OPERATES THROTTLE TENSION LOCK
RETURNS HAND TO ARM REST	R	1	.015	.50	.030	1	L	RETURNS HAND TO ARM REST
ADJUSTS ELEVATOR TRIM	R	2	.072	.55	.032	2	L	OPERATES R.P.M. TENSION LOCK
					.028	2	L	ADJUSTS R.P.M.
RETURNS HAND TO CONTROL COLUMN	R	1	.063	.60				
ADJUSTS ELEVATOR TRIM	R	2	.032	.65				
OPERATES THROTTLE	R	2	.060	.70	.280	1	L	RETURNS HAND TO ARM REST IDLE
RETURNS TO CONTROL COLUMN	R	1	.018	.75				
ADJUSTS ELEVATOR TRIM	R	2	.120	.80				
				.85				
				.90	.065	3	L	TURNS TOWARD PILOT AND REACHES FOR GEAR HANDLE
				.95	.025	3	L	ACTION COMPLETED
					.017	3	L	MOVES HAND TO FLAP CONTROL
				1.00				
RETURNS HAND TO THROTTLE OPERATES THROTTLE	R	2	368	1.05				
				1.10				
				1.15				
				1.20				
								OPERATES FLAP CONTROL RETURNS TO ARM REST IDLE
PLACES HAND OVER THROTTLE KNOBS AND GRASPS ALL FOUR TO HOLD SECURELY IN OFF POSITION	R	2	.145	1.25				
				1.30				
				1.35				
				1.40				
RETURNS HAND TO CONTROL COLUMN	R	1	.095	1.45				
				1.50				
ON THE DECK - PILOT REACHES FOR THROTTLE AND HOLDS	R	2		1.55				
				1.60				
				1.65				

On the definition of links. A link is any sequence of use of two instruments or any sequence of action. It is a connection between (a) a man and a machine, (b) two men, or (c) two parts of a machine. For example, if an operator has to use a telephone, this is a man-machine link. When operator A goes to talk to operator B (or give operator B a piece of paper) this establishes a man-man link. When an operator first twists knob C and then operates switch D, this identifies a linkage between C and D.

A link analysis of a combat information center. An illustration of this procedure applied to a real problem is a study of the Combat Information Center (CIC) aboard the *U.S.S. Louisville.* The original layout of this CIC is shown in Figure 13. As you can see, this is a very complex assemblage of men and machines. It is not really necessary for an understanding of the technique of link analysis to know what all of those equipments represent or to know what all these men do. You will notice from a study of Figure 13, however, that a great many men (plotters, radar operators, and radio men) remain at their posts during an air attack. A few other men (six in all) must move about to carry out their duties. Their paths of movement are shown by lines in Figure 13. The most important thing to emphasize about these paths of movement is that there is no fixed order in which they occur. The actions of any particular officer depend to a large extent on the way the battle goes. About all one can predict about these paths of movement is that they will occur with certain relative frequencies.

Table 2 shows link values between the critical men and equipments in this layout. Along the left-hand margin of the table are the six men who are required to move about when the ship is under air attack. Underneath are shown eight pieces of equipment which these men must use. The entries in the table show the strengths of the various linkages between men and men and between men and machines. Let us examine a few of these to see what they mean.

The link values in Table 2 range from 1 to 9 and are indications of the relative importance of the various links. A link value of 1 means a linkage which is used infrequently and is relatively unimportant. A value of 9 represents a linkage which occurs often and which is very important. Thus, a link value of 6 between the communications officer and evaluator means that these two men must talk to each other with

---

FIG. 12, opposite. Process chart showing activities of a pilot and co-pilot in landing an R5D aircraft. (After Channell.[18])

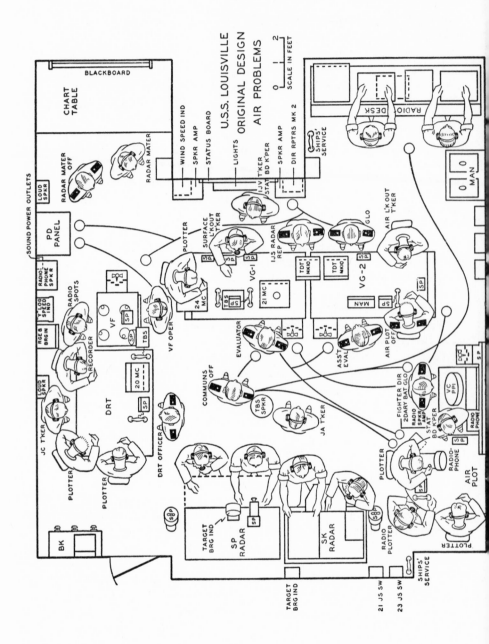

U.S.S. LOUISVILLE
ORIGINAL DESIGN
AIR PROBLEMS

fairly high frequency. A link value of 1 between the communications officer and radio desk means that this officer must go to that station only occasionally and that this is a relatively weak linkage. The link value of 9 between the evaluator and air plot indicates that this man must consult the display on the air plot often and that it is vital for him to do so.

Obtaining the link values. The link values referred to in this problem are *use-importance* values, that is, they give some indication of both the *frequency* with which each link is used and the *importance* of each link. In some instances link values may reflect only the frequency with which various links are used or only their importance.

The best way to obtain information for computing *use-links* is to operate the system in a real or simulated situation and to count the number of times each link is used. From such counts the investigator can then assign simple integer numbers—for example, 1 to 5—to each link according to its frequency of usage. If the system has not yet been constructed, it is necessary to have an experienced person—one who knows how the system will perform or who is familiar with a system like the contemplated one—rate the frequency with which various linkages will probably be used. One way to obtain *importance* ratings is to have experienced persons judge the importance of each linkage.

Use-importance values (like those in this example) are usually obtained by simply adding the use-ranks to the importance-ranks. Although there are many possible mathematical variations of this procedure, there is no theoretical rationale for getting any more complicated. In addition, simple ranks seem to work about as well as more refined link values for most practical problems.

The schematic link diagram. Figure 14 is a schematic representation of the same data shown in Figure 13 and Table 2. Only the critical items of equipment and the critical men are shown in the diagram. They are shown in approximately their correct spatial relationships. Shown also are the link values between men and men and between men and machines. A diagram like Figure 14 shows immediately why this is not the most efficient arrangement of men and equipments. Notice, for example, that to get to the radio desk the communications officer must (a) travel a long way and (b) cross the paths between the evalua-

---

FIG. 13, opposite. Original layout of the combat information center aboard the *U.S.S. Louisville.* The men with shoulder boards are officers; all others are enlisted men. Lines show the movements made by certain men during air attacks.[6]

TABLE 2. *Linkages between men and men, and between men and machines, in the* CIC *of the* U.S.S. Louisville *during air attacks (the problem, men, and equipment are genuine, but the entries in this table are fictitious)*

		Men					
		Com. Off.	Eval.	Asst. Eval.	GLO	Fight. Dir.	VF Oper.
Men	Communications Officer (Com. Off.)		6			6	
	Evaluator (Eval.)	6		5		7	
	Assistant Evaluator (Asst. Eval.)		5				
	Gunnery Liaison Officer (GLO)						
	Fighter Director (Fight. Dir.)	6	7				
	VF Radar Operator (VF Oper.)						
Machines	VC Radar					8	
	Air Plot	1	9				
	Radio Desk	1					
	PD Panel					2	1
	VG Radar No. 1		9				
	VG Radar No. 2			8	7	4	
	VF Radar					3	9
	Director Repeaters					7	

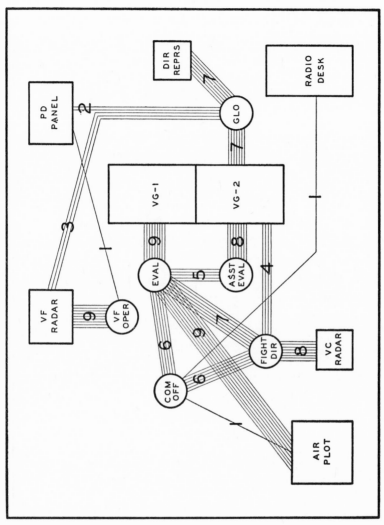

FIG. 14. A schematic representation of the layout of the CIC aboard the *U.S.S. Louisville.* This makes use of the data in Figure 13 and Table 2.

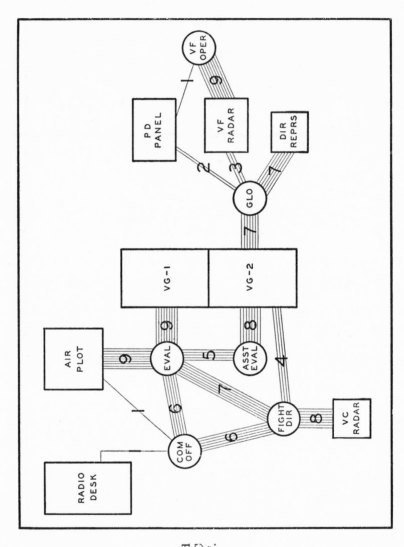

FIG. 15. Revised layout of the CIC for the *U.S.S. Louisville.*

tor and air plot, between the evaluator and fighter director, and between the fighter director and the VG-2. Notice also that the gunnery liaison officer has a long way to go to the VF radar and to the PD panel, which he must consult every once in a while.

An improved layout. A relatively simple kind of analysis of this problem yields a solution illustrated in Figure 15. There you can see that the pathways between men and men or between men and machines never cross. This immediately reduces the possibility of men bumping into each other in the course of their work. Moreover, there has been a substantial reduction in the total length of the paths which the various men must walk in carrying out their duties.

Installing the revised layout. The human engineer's task is not finished when he has produced a layout like that shown in Figure 15. He must still fit the revised layout into the space actually available. Men and equipment must be positioned to avoid pillars and stanchions; to allow free access to doors, elevators and corridors; and to make full use of recessed areas, corners, and cubicles. A good way of doing this fitting job is to use a floor plan of the space and scale cut-outs representing the men and various items of equipment. These can then be moved around on the floor plan until all are accommodated.

During this fitting procedure, think of the layout in Figure 15 as a loose-jointed arrangement with flexible linkages between men and equipment. Men and equipment can be moved around as long as the approximate relationships in Figure 15 are maintained and as long as none of the paths are made to cross each other. You can readily see that this gives you considerable leeway in making up the final arrangement. For example, the radio desk can be rotated 90° and placed along the same wall as the air plot. If there is no space for the radio desk there, perhaps it can be moved down directly opposite and to the left (on the diagram) of the communications officer. The VC radar can be moved to a number of alternative positions without complicating the basic picture. For example, it might be placed near the VG radar, to the left (on the diagram) of the fighter director, or anywhere along a semicircle between these two locations around below the fighter director officer.

Criteria for evaluating layouts. When the final arrangement of men and machines has been fitted into the available space, the investigator should make certain tests to see whether he has actually improved the situation. Three criteria which have been used for this purpose are:

1. AN INDEX OF WALKING. The paths of movements used by the

personnel in carrying out their duties are totalled for a number of typical work exercises or problems, or for a typical length of work period.

2. AN INDEX OF CROWDING. A weighted scoring system is used to determine the amount of interference introduced by crowding. A rating of 3 is assigned to an operator who has completely insufficient space, 2 to an operator whose body is restricted, 1 to an operator who is partly restricted, and 0 to an operator who has adequate space. The number of people affected in each of these ways is multiplied by the appropriate rating and the products are summed across all people affected. The higher the final score, the greater the amount of crowding.

TABLE 3.   *Accessibility ratings based on the width of entries to work areas*

Rating	Width of entry in inches	Explanation
3	Over 36	Room for two men abreast
2	24–36	Free access for one man
1	18–24	Limited access for one man
0	Below 18	Difficult entry

3. AN ACCESSIBILITY SCORE. The passage space available for each person to enter the room and reach his station is determined from physical measurements of the width of the passageways. These measurements are then converted to ratings by means of Table 3. Each rating is multiplied by the number of people affected and averaged across people. An average of 2.5 to 3 indicates excellent accessibility; 1.5 to 2.5 good; 0.5 to 1.5 fair; and 0 to 0.5 poor accessibility.

The illustrations in Figures 13, 14, and 15 have been modified slightly from the real facts of the case as they actually existed in this particular problem. The modifications were primarily to emphasize certain points of technique. In the case of the original design of the CIC aboard this ship, the total length of the paths which men had to walk during an air attack was 62 feet. In the revised design the total length of walking was cut exactly in half—to 31 feet. In the original design the general accessibility rating was only fair—for example, the communications officer could not very easily get to the radio desk. In the modified design the accessibility rating for the layout was excellent.

Link analysis of eye movements. A much simpler type of link diagram occurs in a study by Fitts, Jones, and Milton,[52] who were concerned with eye movements made by aircraft pilots during instrument landing approaches. They installed a motion-picture camera in an airplane so that they could record the sequence of eye movements made by pilots during actual landing operations. These pictures were then compared against a carefully calibrated set of pictures made at the start of each run. This comparison enabled them to determine at what the pilot was

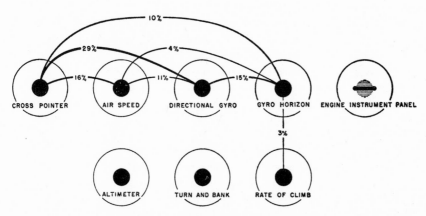

FIG. 16. Percentages of times pilots made eye movements between various cockpit instruments during blind-landing approaches using the Instrument Landing Approach System. Values less than 2 per cent have been omitted. (After Fitts, Jones, and Milton.[52])

looking at any instant. The results of the study are shown in Figure 16. There it is apparent that 29 per cent of all eye movements were in shifting from the cross pointer to the directional gyro—across an intermediate instrument. Ten per cent of the eye movements were in shifting from the cross pointer to the gyro horizon—across two intermediate instruments.

From such an analysis we can now get some ideas for further testing. To take an example, it might be advantageous to reduce the lengths of these links by putting the cross pointer where the turn-and-bank indicator is located. This new arrangement could then be studied to see if it resulted in any improvement over the old one. There is, of course, the possibility that such a redesign might actually result in poorer performance—pilots might have more difficulty locating the correct instrument in the new arrangement, or the new arrangement might be inferior for flight maneuvers other than landings. At any rate, the chief value of this link analysis is not that it provides final answers, but rather that it suggests ways of making new arrangements which can be tested.

MICROMOTION TECHNIQUES

Micromotion study was developed by time-and-motion engineers as a way of making very detailed analyses of short-cycle, repetitive jobs involving mainly hand motions. It has limited usefulness in human-engineering work because it looks at human behavior through a microscope, as it were, whereas most human-engineering problems are macroscopic in proportions. There are, however, some cases where the human engineer may be straining to save seconds in repetitive operations. For these situations micromotion study may be of some value.

Micromotion studies are ordinarily carried out by taking motion pictures of the job to be studied. These motion pictures are taken either with clock readings in the picture or at highly standardized speeds of photography. The investigator then runs the picture over and over again, watching for each element of the work cycle, timing it down to the hundredth of a second, and constructing a detailed analysis of the job.

Therbligs. To aid in such reconstructions, industrial engineers have developed units or fundamental elements of the work cycle. These elements are basic components of the movement pattern. They are essentially names for different activities of the hands, corresponding roughly to the coarser kinds of activities listed in Figure 3. The engineers call these movement elements "therbligs." (This strange sounding term is "Gilbreth" spelled backward but with the "th" unreversed. It was Frank B. and Lillian M. Gilbreth who many years ago pioneered this approach to motion economy.)

There are, of course, an indefinite number of particular movements or elements, depending on the operation, but for most practical purposes industrial engineers have found that seventeen basic therbligs will do. Table 4 lists these seventeen therbligs. One significant thing about this list of therbligs is that it has been standardized. In addition, the therbligs have a generally accepted set of initials and hieroglyphics (not given in the table) which mean the same thing to all those trained in the field. Indeed, when the results of a micromotion analysis are plotted (as in Figure 17), time-and-motion engineers use a standard set of colors to represent the therbligs graphically. To avoid any confusion between the colors, they even specify the manufacturer's numbers of the commercial pencils. The colors and pencil numbers for each therblig are also listed in Table 4.

Micromotion analysis of radar plotting. Figure 17 shows the results of a micromotion analysis of the hand movements made by a radar operator in determining the speed of a target on an intercept plot. An intercept plot is a kind of circular map showing the location of air and surface targets around some central station (in this case a ship). The successive positions of a target (as called in by spotters or radar operators) are usually plotted as X's and are connected by straight lines to show its direction of movement. A series of such connected X's is often referred to as a *track* of the target. Speed was calculated by placing a pencil over the track of the target, measuring the distance traveled by the target in three minutes and then multiplying the distance traveled by twenty to obtain the speed in knots. The distance traveled in three minutes was found by placing the end of a pencil at the center of an X representing one position of the target, and the thumb of the right hand at that point on the pencil corresponding to the plotted position of the target three minutes later. The pencil, with the thumb in place, was then moved to a calibrated scale for measurement.

The chart in Figure 17 shows this operation in therblig terms. It is called a simo-chart because it shows simultaneously the activities of the right and left hands. Notice that the left hand was idle throughout this entire operation. The right hand carried the pencil to a particular raid (raid 12) and positioned the pencil. These are represented by the *transport-loaded* and *position* therbligs. The end of the pencil was then held at one particular X—the *hold* therblig. And so on. The final step of the operation occurred when the operator recorded the speed as he measured it—the *use* therblig at the end.

Roughly a 33 per cent reduction in time was obtained in this operation by designing a special ruler calibrated directly in knots (nautical miles per hour). To calculate speed with this device the radar operator simply aligned the zero marker with one of the plotted points, and then read off speed directly in knots from the ruler at that point representing the position of the aircraft three minutes later.

Exploiting the micromotion analysis. Having prepared a micromotion

TABLE 4. *The therblig definitions and symbols (from Mundel[85])*

Color group and general characteristics	Therblig	Symbol	Color	Eagle pencil*	Dixon Thinex pencil	Definition
	Grasp	G	Lake red	744	369	Begins when hand or body member touches an object. Consists of gaining control of an object. Ends when control is gained.
	Position	P	Blue	741	376	Begins when hand or body member causes part to begin to line up or locate. Consists of hand or body member causing part to line up, orient, or change position. Ends when body member has part lined up.
RED–BLUE— TERMINAL THERBLIGS	Pre-position	PP	Sky blue	740½	418	Same as position except used when line up is previous to use of part or tool in another place.
	Use	U	Purple	742½	396	Begins when hand or body member actually begins to manipulate tool or control. Consists of applying tool or manipulating control. Ends when hand or body member ceases to manipulate tool or control.

RED-BLUE—
TERMINAL
THERBLIGS
(cont.)

Assemble	A	Heavy violet	742	377	Begins when the hand or body member causes parts to begin to go together. Consists of actual assembly of parts. Ends when hand or body member has caused parts to go together.
Disassemble	DA	Light violet	742	422	Begins when hand or body member causes parts that were integral to begin to separate. Consists of taking objects apart. Ends when hand or body member has caused complete separation.
Release load	RL	Carmine red	745	383	Begins when hand or body member begins to relax control of object. Consists of letting go of an object. Ends when hand or body member has lost contact with object.

* See footnote at end of table.

TABLE 4 (CONT.)

Color group and general characteristics	Therblig	Symbol	Color	Eagle pencil*	Dixon Thinex pencil	Definition
	Transport empty	TE	Olive green	739½	391	Begins when hand or body member begins to move without load. Consists of reaching for something. Ends when hand or body member touches part or stops moving.
GREEN— GROSS MOVEMENT THERBLIGS	Transport loaded	TL	Grass green	738	416	Begins when hand or body member begins to move with an object. Consists of hand or body member changing location of an object. Ends when hand or body member carrying object arrives at general destination or movement ceases.
GRAY-BLACK— HESITANT MOVEMENT THERBLIGS	Search	SH	Black	747	379	Begins when hand or body member gropes or hunts for part. Consists of attempting to find an object. Ends when hand or body member has found location of object.

GRAY-BLACK— HESITANT MOVEMENT THERBLIGS (cont.)					
Select	ST	Light gray	734½	399	Begins when hand or body member touches several objects. Consists of locating an individual object from a group. Ends when the hand or body member has located individual object.
Hold	H	Gold ochre	735	388	Begins when movement of part or object, which hand or body member has under control, ceases. Consists of holding an object in a fixed position and location. Ends with any movement.
YELLOW-ORANGE— DELAY THERBLIGS					
Unavoidable delay	UD	Yellow ochre	736	412	Begins when hand or body member is idle. Consists of a delay for other body member or machine when delay is part of method. Ends when the hand or body member begins any work.

* See footnote at end of table.

TABLE 4 (CONT.)

Color group and general characteristics	Therblig	Symbol	Color	Eagle pencil*	Dixon Thinex pencil	Definition
	Avoidable delay	AD	Lemon yellow	735½	374	Begins when hand or body member deviates from standard method. Consists of some movement or idleness not part of method. Ends when hand or body member returns to standard routine.
YELLOW-ORANGE— DELAY THERBLIGS (cont.)	Rest for over- coming fatigue	R	Orange	737	372	Begins when hand or body member is idle. Consists of idleness which is part of cycle and necessary to overcome fatigue from previous work. Ends when hand or body member is able to work again.

Plan	PN	Brown	746	378	Begins when hand or body members are idle or making random movements while worker decides on course of action. Consists of determining a course of action. Ends when course of action is determined.
BROWN— ACCOMPANIED BY THINKING Inspect	I	Burnt ochre	745½	398	Begins when hand or body member begins to feel or view an object. Consists of determining a quality of an object. Ends when hand or body member has felt or seen an object.

* The colors of some of these pencils vary somewhat from the standard colors. They have been selected to match the standard as closely as commercial pencil colors allow.

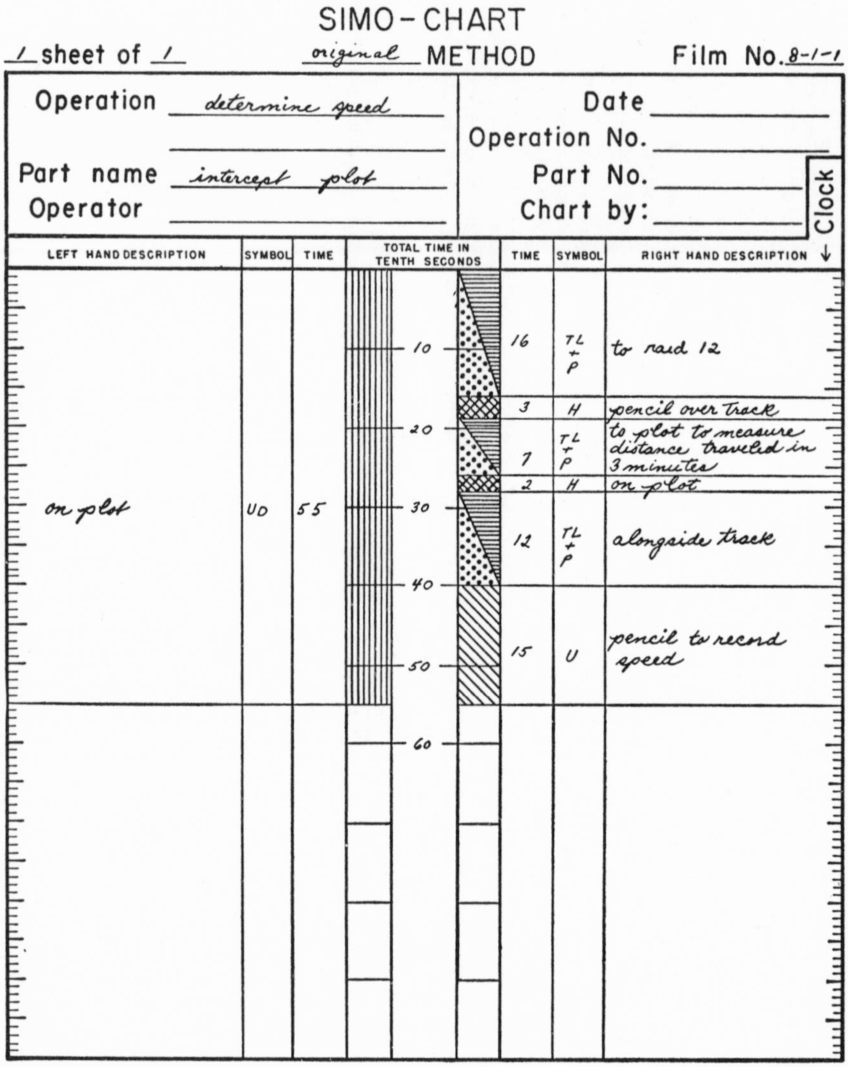

SIMO – CHART

<u>_/_</u> sheet of <u>_/_</u>   <u>*original*</u> METHOD   Film No. <u>*8-/-/*</u>

Operation	*determine speed*	Date	
		Operation No.	
Part name	*intercept plot*	Part No.	Clock
Operator		Chart by:	

LEFT HAND DESCRIPTION	SYMBOL	TIME	TOTAL TIME IN TENTH SECONDS	TIME	SYMBOL	RIGHT HAND DESCRIPTION ↓
			10	16	TL + P	*to raid 12*
				3	H	*pencil over track*
			20	7	TL + P	*to plot to measure distance traveled in 3 minutes*
				2	H	*on plot*
on plot	UD	55	30	12	TL + P	*alongside track*
			40			
			50	15	U	*pencil to record speed*
			60			

FIG. 17. A simo-chart showing a micro-motion analysis of the hand movements made by a radar operator in determining the speed of a target on an intercept plot.[7]

analysis of an operation the investigator typically looks for ways to (a) eliminate useless motions, (b) redistribute the work load among the hands and feet, or (c) provide additional tools or aids to simplify the operator's task. He then develops alternative ways of performing the operation, sets them up for an operator to carry out, photographs and reanalyzes the changed operations, and determines which of the several possible ways will enable operators to do the job fastest.

A CRITIQUE OF THE
OBSERVATIONAL METHODS

For all their usefulness in studying man-machine systems, the observational methods have some serious limitations. Out of these limitations, however, come some positive suggestions of importance:

Observation may affect the behavior being studied. The first drawback of direct observation is that the process of observing may itself influence the things you are studying. There is little doubt, for example, that the mere presence of a time-and-motion engineer affects the rate at which workers produce and the way in which they do it. In the link analysis of the eye movements of pilots cited above, the recording process required that the pilots wear a special hood and that the instrument panel have a special mirror in which eye movements were reflected. These obviously changed the nature of the job and undoubtedly affected the pilot's performance. Occasionally, such interferences may be so great that it would actually be dangerous to conduct the study.

The positive principle to come from this is that in human studies it is important to keep the method of observation from influencing the behavior under study. By this I mean more than simple physical interference with the subject's movements or actions; I mean also the more subtle influences exerted on the subject's behavior by the knowledge that he is being studied. People do not act naturally when they know they are being observed. Although this is a very hard condition to satisfy, it is a very important one when observational methods are used. It is for this reason that psychologists often use one-way vision mirrors, concealed microphones, and other recording subterfuges to avoid the appearances of an artificial laboratory situation.

The difficulty of finding significant relationships. The observational methods frequently yield pooled data from which it is difficult or impossible to tease out significant relationships. The service-observing of long-distance dialing is a good case in point. The results of this work pointed to a trouble spot, but they left unanswered many important questions which are really at the root of the problem: Who makes these errors? Are they due only to a few operators, or are they common to all operators? Do experienced operators make fewer errors than inexperienced ones? Do the errors increase at certain times of day? When a wrong key is hit, what was the operator supposed to have hit? Are some letter and number combinations more troublesome than others? And so on.

Once the investigator has some ideas about the factors which are important in the task, he should look for ways of setting up conditions of observation so that some of his questions can be answered. For example, separate observations can be made on some experienced and inexperienced operators to get at this factor. Observations can be made systematically at various times of day, and so on.

Sometimes, however, it is not possible to set up conditions of observation to isolate the factors that influence the behavior in question. For example, statistical studies show that accidents occur at night more frequently than one should expect on the basis of the amount of traffic involved. Why is this? Among many possible explanations are these two: (1) Darkness reduces visibility and so increases the likelihood of accidents; (2) Vehicular traffic at night is qualitatively different from that during daylight hours (relatively fewer local delivery vehicles operate at night, for example). To separate these effects we would have to compare (a) the number of *night* accidents which occur with *daylight* traffic conditions against (b) the number of *night* accidents which occur with *nighttime* traffic conditions. The trouble is that these two variables go together—when night falls we get both reduced visibility and a change in the character of the traffic. There is no way to untangle the two factors because we have no control over the situation.

To sum up, then, we can say that the observational methods can serve as a fruitful source of hypotheses and hunches. Once these hypotheses and hunches have emerged, observations should be made under conditions which systematically sample the factors of interest, if this is possible. If not, they should be tested in experiments designed for the purpose.

What should be observed? Another disadvantage of the observational methods is that the investigator may frequently collect data on irrelevant operations because he sometimes does not know in advance what is important. In short, he may end up with a great deal of useless data.

The positive suggestion in this case is that it is well worth spending some time thinking about what it is you want to observe and why, before you start the observations. Further, it may be useful to take stock after the first day's observations. Do these seem to tell you something interesting? Is it likely that something important will come of them? Are they worth going on with?

Some observations are costly. Directly associated with the above is the limitation that some observational studies can be extremely costly and time consuming. In the link analysis of the pilot's eye movements, for example, data were reported on only 160 minutes of recording, but this meant the detailed study of 76,800 frames of photography requiring many man-months of analysis!

How general are the findings? One of the most serious limitations of the observational techniques is that they frequently do not tell us anything which we can generalize beyond the particular system under study at this moment. In the study of a radar system, all of our observations may frequently be thrown out the window the minute a new radar is designed—which happens often enough to be discouraging. In many businesses, tests are continually being run on one particular kind of equipment, or gadget, or layout, versus another. The trouble is that new gadgets are always coming out, and each new one means another study. In the long run, the most economical kind of study is not the observation of particular systems, but rather the basic investigation of functional relationships between human and physical variables. The issue is one of specific product testing or evaluation versus basic research.

The difficulties of evaluating changes. Perhaps the most serious criticism of the observational methods is that they are usually used with some extremely simple, and extremely risky, procedures for evaluating improvements. In conventional activity analysis, process analysis, or micromotion study, the investigator observes a process, times it, and charts it. From a study of these observations he often comes up with ideas for improving the process. These improvements are then set up in a new process and the whole series of observations is repeated. If the new set of observations shows that the revised process takes less time, or is more efficient, it is assumed to be better. Although this kind of simple before-and-after testing is easy to do, and is frequently very convincing, it is risky and may lead to fallacious conclusions.

TABLE 5. *A summary of the experimental conditions and findings of the Hawthorne experiments on relay assembly (from Roethlisberger and Dickson[95])*

Period	Length in weeks	Experimental conditions of work	Length of work-week in hours	Percent increase in output over period 1
1	2	Standard	48	
2	5	Operators moved from regular department to test room	48	15.2
3	8	Standard but with a change in piece rate method of payment	48	19.1
4	5	Two 5-min. rest pauses, one in the morning, the other in the afternoon	47:05	16.6
5	4	Two 10-min. rest pauses, one in the morning, the other in the afternoon	46:10	28.5
6	4	Six 5-min. rest pauses, three in the morning, the other three in the afternoon	45:15	26.3
7	11	15-min. rest pause and lunch in the morning plus 10-min. rest in the afternoon	45:40	19.1
8	7	Same as period 7, but with 4:30 P.M. quitting time	43:10	31.4
9	4	Same as period 7, with 4:00 P.M. quitting time	40:40	29.6
10	12	Same as period 7	45:40	39.3
11	9	Same as period 7, but with no work on Saturday mornings	41:40	27.4
12	12	Same as period 3	48	44.0
13	31	Same as period 7	45:40	49.8

One of the most impressive illustrations of the dangers of this kind of testing is the now classical study of relay assembly made at the Hawthorne Works of the Western Electric Company. In this study five girls were placed in a test room and their output was studied over a two-year period. A summary of the experimental conditions of work and the findings of this study are shown in Table 5. During Period 1, which lasted for two weeks, the girls were studied at their usual workplaces and under their ordinary conditions of work. In Period 2, which lasted for five weeks, the girls were studied immediately after being moved from their regular department to the experimental room. This simple change produced a 15 per cent increase in production.

Thereafter, a series of variations in rest pauses and working hours was tested. In Period 7, for example, the girls were given a 15-minute rest pause in the morning and a 10-minute rest in the afternoon. This reduced the work week from 48 hours to 45 hours and 40 minutes. Despite this 5 per cent *decrease* in the length of the work week, production *increased* by 19 per cent over the first observation period.

If this study had been terminated at the end of Period 7, the investigators might very well have decided that they had made a definite improvement in efficiency and let it go at that. Fortunately, they were not that gullible. They tested a couple of other conditions, and then, in Period 10, tested the same conditions as they had previously tried in Period 7. This time production increased by 39 per cent. They then tried a couple of other schedules, and, in Period 13, again set up the same conditions as they had tested in Period 7. This time production had increased by about 50 per cent!

Perhaps the safest conclusion to draw from this entire experience is that production seemed to increase gradually throughout the experiment no matter what the investigators did. There is probably no better example in the literature of the danger of simple before-and-after testing.

What was the trouble? The trouble was that there were some extraneous variables—for one, the entire social climate of the experiment—which were contaminating the results. The fact that this was an experiment—and the girls knew it—affected their production. But if these schedules had been instituted in the plant as a whole, there is no guarantee that improvements in production would have resulted. And therein lies the real danger of this kind of testing. It seems to give you an answer, but that answer may not be trustworthy.

What is the moral of this story? It is this: If you want to evaluate possible improvements in procedures, be sure to follow the rules of careful scientific experimentation discussed in Chapter 5. Observation by itself may be misleading. Observation *with* the controls demanded by the experimental scientist can be made much more dependable.

Chapter 3 | **Methods for the study of**

accidents and near-accidents

Just as physicians have been able to discover many important and basic things about human physiology by studying diseases, so also the engineer can discover many important things about man-machine systems by observing malfunctioning ones—that is, by observing what happened, or reconstructing what must have happened, preceding and during accidents. The Civil Aeronautics Administration, for example, has a very elaborate procedure set up for the investigation of all aircraft accidents. From studies like these they have been able to uncover many instances of man-machine friction and to correct the difficulty before it resulted in still other accidents.

BASIC REQUIREMENTS FOR ACCIDENT DATA

For all their value in the analysis of certain man-machine problems, accident data can be more hindrance than help. Indeed it is probably correct to say that there are more useless than useful accident data. If accident data are to serve a useful function they should satisfy at least five minimum requirements.

DEFINITION

The first requirement to be met before you can collect good accident data is that you must have an adequate definition of precisely what constitutes an accident. If an investigator is to go about collecting

76

accident data, he must know exactly what he is going to count. Analyses of existing accident data show that there are several ways of defining accidents, and these definitions are not always kept distinct. For example, legal definitions of accidents, as used in some industries, are generally unsatisfactory for human engineering purposes. The reason for this is that legal definitions are too much concerned with establishing guilt or responsibility. The lawyer is primarily concerned with, "Who pays for it?" and not with, "How could we have prevented it?"

A working definition of accidents. It is hard to give an adequate definition of an accident, and it is probably for this reason that the problem is so often ignored. For human engineering study, the following definition may be useful as a starting point:

> An accident is an unexpected and undesirable event which arises directly from a work situation; that is, from faulty equipment or the inadequate performance of a person. There may or may not be personal injury and damage to equipment or property. Accidents, however, always interrupt the normal work routine and are associated with increased time delays or errors.

Perhaps the most important thing to notice about this definition is that it is broader than most. It covers the gamut of events ranging from near accidents in which there is no personal injury or damage all the way to incidents which may be accompanied by death and extensive property loss. As we shall see later, the investigation of near accidents can be very helpful.

AN ADEQUATE SYSTEM FOR REPORTING ACCIDENTS

Three essentials of a good system for collecting accident data are: proper personnel, a good accident report form, and central-office facilities for handling accident reports.

Personnel. Perhaps the most important thing to say about personnel is that somebody should be delegated to collect accident data. Depending on the size of the organization, this may be a single person or a group of people. But whether one, several, or many people are involved, each one should know that it is his responsibility to report every accident that occurs within his jurisdiction. If the responsibility for accident reporting is unassigned, or if each member of an organization is left to do his own, many accidents will never get into the files or, if they do, will be incompletely or improperly reported.

The second requisite is that personnel who have responsibility for collecting accident data should be trained in their job. They should know what to look for, how to question accident victims and witnesses, and how to fill out the accident report form. Properly trained investigators are much more likely to give complete and accurate reports than are untrained persons. Reports prepared by trained personnel, moreover, tend to be more uniform and more consistent with each other over a period of time.

The final thing to say about personnel is that they should be competent people who take their jobs seriously. Sometimes the job of safety officer is used as a kind of "Siberia"; that is, it is assigned to misfits, or to people who cannot seem to get along in any of the regular departments within an organization. To get good accident reporting done, the personnel assigned to the job should be capable and should recognize their assignment as an important and useful one.

Accident report forms. Carefully prepared and pretested accident report forms should be available (a) to make it easy for the investigator to prepare his accident report and (b) to serve as a constant reminder of the things that should be written up. The use of specially prepared forms also induces uniformity of accident reporting—a factor of some importance when one tries to collate and make sense out of many reports. There is a fuller discussion of the accident report form starting on page 81.

Central-office facilities. Good accident reports are of little use unless there is some place where they can be collected, organized, compiled, and stored for future reference. Last but not least, it should be easy for the investigator to get his report into the central office.

COVERAGE SHOULD BE COMPLETE

All accidents covered by the definition should be reported, or, at the least, any accident covered by the definition should have an equal likelihood of getting into the record. If accidents which happen to drivers of taxicabs are less likely to be reported than accidents which happen to ordinary drivers, the accident statistics will be seriously biased. Similarly, if accidents which happen to a general in the Air Force, the captain of a ship, or the foreman in a factory are less likely to be reported than those which occur to rank-and-file operators, the data will again be biased.

Bias also comes into the data if certain types of accidents are less likely to be reported than other types. In vehicular-accident reporting, for example, fatal and serious-damage accidents are much more likely to be reported than are those involving minor injury and property damage. Since in the great majority of accidents it often appears to be a matter of chance whether a particular accident results in serious or minor damage, this incompleteness of reporting imposes a serious limitation on the investigation of accident causes from the available records.[79] To sum up, then, complete coverage of accidents is necessary to insure that the statistics tell you about accident *occurrence* and not accident *reporting*.

ACCIDENTS SHOULD BE REPORTED FULLY AND ACCURATELY

The previous section was concerned with the problem of accident coverage, that is, of getting *all* accidents reported and into the record. That is primarily a problem of statistical sampling. In this section, our concern is quite different. By full and accurate reporting we mean that *each* accident report should be a complete and thorough description of exactly what happened.

Some difficulties of observation and reporting. At once we run into all the limitations of human observation and recollection about which we spoke earlier. In particular, if accident reporting is done by the people involved in an accident, one may get seriously distorted observations. Few people like to admit they have contributed to an accident. If a person does admit that his own actions contributed to the accident in some way, he is likely to interpret, or edit, his account so that he appears as blameless as possible.

Police statements contain numerous instances of such distortions. Here is one taken from the scientific literature. LeShan and Brame report the case study of a man who had been involved in four separate automobile accidents.[71] In each case he appeared to have been an innocent victim since he said he had been a passenger in the front seat and had "generally been talking to the driver when it happened." In one of these accidents he had hurt his elbow seriously because it had been outside the ventilator window when the accident occurred. This statement seemed odd to the investigator who then probed more thoroughly into the exact circumstances of the accident. After five minutes a completely different picture emerged from that which had been originally reported.

It is true that the narrator had been a passenger in the front seat, but he had decided to clean the windshield. He thrust his hand with a towel through the side ventilator window while the vehicle was traveling at 50 miles per hour. The towel flapped over, covering most of the windshield so that the driver could not see. It was this last event which resulted in the accident.

False reporting. There is also the reverse kind of error to guard against: accidents may be falsely reported. For example, passenger-car drivers occasionally report an accident with a large truck or bus, although no such accident occurred.[79] The unscrupulous person sometimes makes such a complaint to get minor repairs made on his vehicle at someone else's expense. Because of the size of modern trucks it is actually possible for a minor collision to occur without the truck driver feeling or hearing it. This has constituted a sufficiently strong legal argument for passenger car drivers to recover "damages" in some instances.

Related kinds of false accident reporting occur in industry and in the military services. In industry, workers sometimes make such claims to "earn" time off with compensation, or to recover monetary "damages." Such false reports occasionally work because certain types of ailments are difficult to diagnose correctly. Whatever the motivation, the effects of false reporting on accident statistics should be apparent without elaboration.

Some rules for getting accurate accident reports. In the face of difficulties like these what can the investigator do to get full and accurate accident reports? Here are some rules to observe:

1. The investigator should be as objective as possible in collecting the facts (see Chapter 1). Above all, he should avoid "taking sides" with any of the persons involved. He should remember that his job is to collect data as accurately as he can.

2. Investigate promptly. The best way to guard against many of the distortions of human memory is to collect data as soon as possible after an accident has occurred.

3. Be sure to get information from as many witnesses as possible. Witnesses are often likely to be more objective in their reporting than are the people directly involved. Moreover, when data are available from several witnesses the investigator can often spot discrepancies

between different accounts. From these he can then form some estimate about the credibility that can be attached to them and decide if and where he should do more questioning.

4. Interrogate victims and witnesses individually and privately. People are more likely to talk freely to an accident investigator if they are interviewed in this way. This is especially true if what a witness has to say might be interpreted as derogatory to any of the other persons involved. In addition, the investigator is much more likely to find discrepancies if people are questioned individually, before they have had a chance to compare notes.

5. Be sure to probe carefully and fully. The investigator should not be satisfied with the first statement or explanation a person gives.

6. Use an accident report form to be sure that information is collected on all the relevant facts. Some suggestions about accident report forms follow.

Accident report forms. To a considerable extent, accident report forms have to be tailor-made for particular jobs. The kinds of accidents that happen on the highway are quite different from those which occur in the air, in the home, or in the machine shop. For this reason, it is impossible to produce a single form which will be satisfactory for all situations. The items below merely suggest some of the things that should probably appear on most report forms. Obviously, the investigator will want to modify, add to, or subtract from this list according to his own needs.

1. Identification of the persons involved: name, number, crew, shift, age, previous work experience, previous accident record.

2. Time of the accident: hour, day, month, year.

3. Place of the accident: town, plant, place of work, department, specific location in the department.

4. Identification (names and addresses) of all witnesses to the accident. Names of foremen and supervisors.

5. Nature and severity of the injury (if injury is involved): amount of lost time and cost of the injury, name of the attending doctor, and record of treatments.

6. Description and cost of property damage or spoilage.

7. An exact description of the accident. Among the items which can be relevant are the following:

a. A full description of the accident, stating, for example, whether the person fell or was struck, and all the factors contributing to the accident.

b. Identification of the machine, tool, appliance, gas, liquid, or other object which was most closely associated with the accident.

c. If a machine or vehicle was involved, identification of the specific part which was involved; for instance, the gears, pulley, or motor.

d. A judgment about the way in which the machine tool, object, or substance was unsafe.

e. Description of mechanical guards or other safeguards (for example, safety goggles) which were provided.

f. Statement about whether the injured person or persons used the safeguards which were provided.

g. Description of the unsafe action of the injured person which resulted in the accident (for example, lifted weight with bent back, removed safety screen from pulley, did not wear goggles, ran down stairs).

h. The investigator's, witnesses', or victim's best guess as to why the person acted unsafely. (Examples: was indulging in horseplay, became rattled when a minor emergency arose, had never received proper instruction in the use of equipment, was in a hurry.)

i. The investigator's, witnesses', or victim's opinion about ways of preventing further accidents of this type. (Examples: provide better illumination, provide safety goggles, provide better guards, do a better job of instruction).

Some further suggestions about report forms. Report forms should be printed in a convenient arrangement and should, above all, provide *plenty* of space for answers. One of the most common difficulties with report forms in general is that they do not leave the respondent enough room for his answers. There is scarcely anything more frustrating or inhibiting.

Report forms should be pretested. A preliminary form should be made up and tried out for several weeks or months to find out if it does the job well. This will help the investigator discover whether some questions are irrelevant, whether he should ask other questions, and whether still others should be reworded. The final version should be printed only after such an initial trial period.

GOOD ACCIDENT DATA SHOULD SUGGEST CURES

One of the most serious criticisms which can be levelled at many accident statistics is that they fail completely to answer the question:

What was wrong that should have been corrected? For example, a set of accident data collected by a large insurance company lists as the primary causes of street railway accidents such clichés as *faulty attitude, impulsiveness, irresponsibility, slow reaction,* and so on. (Refer to the discussion on page 19 about such terms.) In aircraft accident reporting it has long been the custom to ascribe accidents to such things as *pilot-error, poor judgment, poor sense of spatial relations,* and so on. Although such labels appear to explain behavior, they really do not.

"Pigeon-holing" human actions does not help reduce accidents. One trouble with such labels is that they imply there is such a mythical being as a person who has exactly the correct attitudes, who is never impulsive, who is never irresponsible, or who never makes a mistake in judgment. This is nonsense, of course. Every one of us is guilty of every one of these faults at some time or another. Another trouble with this kind of labelling is that it requires the investigator to make a very tricky and difficult judgment about someone else's mental makeup. Deciding whether someone else has poor attitudes, is impulsive or irresponsible, is something which even well-trained psychologists hesitate to undertake without very extensive study and analysis.

Finally, assigning a name to something is an easy way of avoiding an issue or covering up for ignorance. To say that a series of related accidents was caused by *pilot-error, poor judgment,* or *faulty attitudes* is of little use because it provides us with no cues about ways of eliminating the accidents. How do you eliminate pilot errors? Or how do you correct poor judgment? A much more positive approach to the problem is to redesign the machine, or work situation, so that accidents become less likely to happen. Some illustrations are given below in the section on the critical-incident technique.

A system for tabulating accident data. In attacking the problem of accidents, the human engineer frequently wants to know, "On what kinds of machines do accidents often occur?" "Are there some work situations that seem to be more hazardous than others?" Armed with answers to questions of this sort he then knows which situations or machines should be carefully analyzed for improvement. To get such answers, however, the investigator needs a consistent and useful system for summarizing, compiling, and tabulating accident data.

One workable system for compiling accident data has been published by the American Standards Association.[2] It is the work of many people and its purpose is to provide a statistical method of recording the accident facts which are essential for accident prevention. Emphasis has been placed on those factors which are commonly recognized as being most closely related to accidents and which are of most value in practical preventive work. Another important feature of the system is its flexibility, since it can be adapted to many different kinds of industries and situations. The system has been worked out in such detail that it would take many pages to present it fully. The material below constitutes only a bare outline.

The ASA system for compiling accident statistics uses six major headings:

1. THE AGENCY, defined as the machine, tool, object, or substance which is most closely associated with the injury. The full listing of agencies and agency parts covers over 30 pages in the ASA publication and is designed to cover almost every conceivable kind of industrial device. Examples of agencies are: grinders, ironers, kneaders, lathes, washers, elevators, conveyors, animals, chemicals, dusts, and radiating substances.

2. THE AGENCY PART, defined as the particular part of the agency which was most closely associated with the accident. Examples of agency parts are: belts, pulleys, gears, indicators, gages, flywheels, valves, and gaskets.

3. THE UNSAFE MECHANICAL OR PHYSICAL CONDITION, defined as the condition of the agency which could have been guarded or corrected. Examples are: improperly guarded agencies, defects (sharp edges, decayed, worn, frayed, or cracked parts), hazardous arrangements, improper illumination, improper ventilation, and unsafe dress.

4. THE ACCIDENT TYPE is the way in which the injured person came into contact with the agency or substance. Examples are: striking against; falling; being caught in, on, or between; and inhaling, absorbing, or ingesting.

5. THE UNSAFE ACT is the unsafe behavior which resulted in the accident. Examples are: operating without authority; working at unsafe speeds; distracting, teasing, and abusing; and making safety devices inoperative.

6. THE UNSAFE PERSONAL FACTOR, defined as the mental or bodily characteristic which permitted or resulted in the unsafe act. Examples are: improper attitude, lack of knowledge or skill, and bodily defects.

The sixth heading in this system is probably the least useful for reasons which have already been discussed. However, even that has some useful aspects, since it is important to know if inadequate instruction or bodily defects were in any way associated with the accident.

An example of the use of the ASA system. To illustrate the use of the ASA code consider the following incident:

A worker was injured when his hand was caught in the running end gears on a lathe. Contrary to instructions, the worker had removed the guards to facilitate cleaning.

The coding for this accident is:

Agency: Lathe
Part of agency: Gears
Unsafe mechanical condition: Unguarded
Accident type: Caught in, on, or between
Unsafe act: Removing safety devices
Unsafe personal factor: Disregard of instructions

From these coded facts an investigator can reconstruct the accident as follows: Someone was injured because he was caught between the gears of a lathe which were unguarded. The injured person had removed the safety devices contrary to instructions.

THE STUDY OF NEAR-ACCIDENTS

Some disadvantages of collecting accident data are that (1) you have to wait for an accident to happen before you can count it, (2) accident data are often effective only in large organizations where there are enough accidents of the same kind over a period of time to provide some estimate of frequency, and (3) people are reluctant to report that they have been involved in certain types of accidents. The study of near accidents obviates some of these difficulties. Since near accidents generally occur much more frequently than accidents, even small organizations can often collect a sizeable amount of data in a relatively short time. Moreover, people are more willing to talk about "close calls" than about accidents.

The general rationale behind the collection of near-accident data is that the severity of an accident appears to be largely fortuitous. For example, one investigator found that in a group of 330 accidents of the same kind, 300 resulted in no injury, 29 in minor injuries, and only one in a major injury.[63] This suggests that the real importance of any accident is that it *identifies a situation* which could potentially result in injury or damage. Whether an accident did, or did not, result in injury is much less significant. The important thing is to eliminate an accident situation. The study of near accidents often makes it possible to take remedial action before a single serious accident occurs.

A NEAR-ACCIDENT STUDY IN THE AIR FORCE

Vasilas and his co-workers[108] have developed some procedures for collecting near-accident data which should be useful for human engineering purposes. In particular, they developed a report form for collecting near-accident data in the Air Force, studied some alternative methods of collecting the data, and worked out a system for classifying the reports. There are a number of items of interest in the Vasilas report, but we shall look only at some points of technique.

A comparison of three methods of collecting data. These investigators compared three different methods of collecting near-accident data: (a) an *at-source* method; (b) a *group orientation* method; and (c) a *command* method. In the first method, individual report blanks were made available for reporting hazardous incidents after flight. A blank was attached to the flight record of each aircraft, and a supply of them was located in such places as the BOQ's (Bachelor Officer Quarters), the Operations Office, the lounge, and the hangars. Aircrew members were urged to report any incidents they observed or experienced. The person completing the blank brought it to Operations, where he sealed it in an envelope which was placed in a collection box. The Flight Safety Officer then collected the incidents and sent them to a civilian research organization which was conducting the study.

In the group orientation method, an interviewer (usually the Flight Safety Officer) had groups of about ten aircrewmen appear in his office at certain times. He instructed them concerning the nature of the procedure and the value of collecting such incidents. He then had the men write descriptions of whatever near accidents they could recall. These were then placed in envelopes, sealed by the men, and treated as above.

The command method for collecting data was a system in operation at several air bases. Subordinate units were required to maintain weekly activity reports which include reports of accidents and near accidents.

Results. Nearly two and one-half times as many incidents were obtained with the group orientation method as compared with the at-source method. Some of the reasons for this are perhaps that (a) the presence of the Flight Training Officer put some pressure on the men to recall incidents, (b) the presence of other people made each man more likely to conform, and (c) the knowledge that the meetings were to be held repeatedly made the men more alert to the occurrence of hazardous incidents.

Over seventeen times as many incidents were collected by the group orientation method as were collected with the command method. Most of the incidents reported in the command method were incidents which were almost certainly observed by someone else.

Most interesting, however, was the distribution of types of items. In the command method, most of the reported incidents fell into the category of "mechanical malfunction." In the group orientation method, most of the reported incidents fell in the category of "personal errors."

A reasonable interpretation of these results is that crew members hesitate to report incidents under the command method because they are afraid that the incidents may be used to evaluate or punish them. In fact, incidents of this sort are apparently used for this purpose at some bases. If large numbers of incidents are to be collected it seems likely that the people involved must have guarantees of anonymity. It may well be that the best way to enlist the co-operation of military personnel is to have the data collected by an outside civilian agency, as was done in the Vasilas study. This is one way of insuring that military commanders do not misappropriate research data for purposes of personnel evaluation.

Summary and recommendations. Let us review the three major points of interest in this study. Some of them merely strengthen points which have already been discussed earlier in this chapter, but they are worth repeating because of their importance.

1. The kind and amount of accident data you collect depends on the system which has been set up for reporting accidents. In this study one system yielded over seventeen times as many incidents as another. More important is the fact that different kinds of accidents were reported with the different systems.

2. It is better to have someone responsible for collecting accident data than to allow everyone to do his own reporting.

3. If you want to collect the maximum amount of accident data, you have to divorce the threat of punishment or evaluation from accident reporting. One way to do this is to guarantee anonymity for the

persons reporting accidents. Another is to have accident data collected by an independent research agency which clearly states its research goals at the outset.

NEAR-ACCIDENTS IN LONG-HAUL TRUCK OPERATIONS

For another illustration of the use of near accidents refer to a study by McFarland and Moseley (pp. 240 ff.),[80] who used it to find out what types of near accidents occur in long-haul trucking operations, their frequency, and the conditions under which they take place. The data were collected by a trained observer, a former truck driver holding a union card. This man accompanied 17 drivers on 20 trips totaling approximately 5,000 miles, and looked for emergency situations which could easily have led to an accident. Whenever such an incident occurred the observer diagrammed clearly the course of action taken by the driver of each vehicle and attempted to define the errors which led to the critical situation. In addition, the observer made additional notes on the illumination, time of day, weather, visibility, type of road, elapsed time of the incident, and the speeds of the vehicles involved.

Although this study turned up a number of interesting findings, we shall confine our attention to those which concern the technique itself. The most important point is that this method yielded many more items than could be obtained by waiting for real and serious accidents. In all, 48 incidents, or about one in every 100 miles, were reported. It is, as noted above, an efficient way to collect data.

The two principal disadvantages of this method are that (a) the definition of a near accident depends on the judgment of the observer and that (b) the presence of an observer in the vehicle has a tendency to put the driver on his "good behavior." The first can be controlled to some extent by training the observer; the second is much harder to do anything about.

THE CRITICAL-INCIDENT TECHNIQUE

A special form of accident study is an adaptation of a method which has been applied to job analysis, the analysis of performance requirements, and proficiency testing.[53] It is called the critical-incident technique.

The critical-incident technique has been used to collect both accident and near-accident data without any discrimination being made between the two types of data. However, in particular cases the investigator may confine his attention to one or the other type of data. In essence, the investigator interviews or otherwise questions a large number of people who have used a particular instrument, machine, or vehicle, and asks them: "Tell me about some error or mistake you have made in using this particular machine." The basic assumption underlying the method is that from a large number of such incidents one can discover difficulties which are critical; that is, which led to or might have led to a crisis or accident.

To be useful the incidents must be detailed enough (a) to allow the investigator to make inferences and predictions about the behavior of the person involved and (b) to leave little doubt about the consequences of the behavior and the effects of the incident.

AN ILLUSTRATION OF THE CRITICAL-INCIDENT TECHNIQUE

Fitts and Jones used the critical-incident technique to good purpose in surveying psychological problems relating to the use and operation of aircraft equipment.[50, 51] They asked a large number of pilots if they had ever made, or had seen anyone else make, "an error in reading or interpreting an aircraft instrument, detecting a signal, or understanding instructions." They checked very carefully to be sure that the men were actually present when the events occurred. We can get some idea about the kinds of stories they got from these verbatim accounts:

1. "It was an extremely dark night. My co-pilot was at the controls. I gave him instructions to take the ship, a B-25, into the traffic pattern and land. He began letting down from an altitude of 4,000 feet. At 1,000 feet above the ground I expected him to level off. Instead, he kept right on letting down until I finally had to take over. His trouble was that he had misread the altimeter by 1,000 feet. The incident might seem extremely stupid, but it was not the first time that I have seen it happen. Pilots are pushing up plenty of daisies today because they read their altimeter wrong while letting down on dark nights."
2. "I was flying at 25,000 feet in a P-47 on my first combat mission, but had mistakenly read the hands on my altimeter and was

TABLE 6.   *Classification of 227 "pilot-error" experiences collected by Fitts and Jones*[51]

Type of error	Number of errors
I. Errors in interpreting multi-revolution instruments:	
A. Errors involving an instrument which has more than one pointer, e.g., misreading the altimeter by 1,000 feet, the clock by 1 hour, etc..	40
B. Errors involving an instrument which has a pointer and a rotating dial viewed through a "window," e.g., misreading the tachometer by 1,000 rpm, the air-speed meter by 100 mph	8
II. Reversal errors, e.g., reversals in interpreting the direction of bank shown by a flight indicator, reversals in interpreting direction from compasses, etc.	47
III. Legibility errors:	
A. Instrument markings difficult or impossible to read because of improper lighting, dirt, grease, worn markings, vibration, or obstructions	32
B. Parallax: Difficulty in reading an instrument because of the angle at which it is viewed	5
IV. Substitution errors:	
A. Mistaking one instrument for another, e.g., confusing manifold-pressure gauge with tachometer, clock with air-speed meter, etc.	24
B. Confusing which engine is referred to by an instrument	6
C. Difficulty in locating an instrument because of unfamiliar arrangement of instruments	6
V. Using an instrument that is inoperative, i.e., reading an instrument which is not working or is working incorrectly	25
VI. Scale interpretation errors, i.e., errors in interpolating between scale markers or in interpreting a numbered graduation correctly	15
VII. Errors due to illusions: Faulty interpretation of the position of an aircraft because body sensations do not agree with what the instruments show	14
VIII. Signal interpretation errors: Failure to notice a warning light in the aircraft, or confusing one warning light with another	5

under the impression that I was at 35,000 feet. I called in some unidentified aircraft which were level with our formation, and, consequently, actually were at 25,000 feet. Since I mistakenly reported them at 35,000 feet, they were assumed to be enemy aircraft. A good deal of confusion resulted. I believe some improvements can be made in our present altimeter."

3. "I glanced away from the instruments while making a steep bank in a C-47. Upon glancing back at the artificial horizon, I was confused as to the direction of turn shown by the little pointer which indicates degree of bank. Upon beginning to roll out, I used exactly opposite aileron control from what I should and thereby increased the bank to such an extent that it was almost 90° and considerably dangerous."

If these were isolated incidents, the stories would probably not have been very helpful. But, as it turned out, Fitts and Jones were able to collect 270 such "pilot-error" incidents and find many points of similarity in them. Table 6 is condensed from their report, and it shows some of the kinds of errors they found.

There are a number of interesting things about the data in Table 6, but the one which concerns us most is the philosophy of this approach as compared with that adopted in the study of street railway accidents mentioned earlier. It would have been easy to lump almost all the incidents in Table 6 under such labels as *carelessness, inattention, pilot-error, inadequate training*, and so on. In fact, this has been the usual method for disposing of human errors. Instead of taking us into this blind alley, however, the classification of errors in Table 6 suggests some remedies. The fact that so many pilots report serious errors in reading multi-revolution indicators suggests that perhaps such an instrument is too hard to read. This, in fact, led to some experiments on the design of long-scale indicators and some recommendations for a new kind of altimeter which is less confusing than the conventional one.[61] This redesigned altimeter has recently been put into production by an instrument company.

To take just one other illustration, the reversal errors shown in Table 6 might, in the more usual way of dealing with them, have simply been dumped into a bin called *inadequate sense of direction*. But here again, the fact that so many pilots report difficulties of this sort suggests an entirely different approach. Perhaps pilots make these errors because the directional relationships displayed on some instruments are confusing. This has led to a whole series of experiments on the design of attitude indicators, one of which we cited earlier.[75] Many instruments in aircraft are, in fact, confusing, but what is more important, their confusability can be reduced by proper design.

AN EVALUATION OF THE CRITICAL-INCIDENT TECHNIQUE

Partly because it is ordinarily used to gather both near-accident and accident data, the critical-incident technique is an efficient way of collecting information about sources of man-machine friction. This kind of study can be done by mail, it does not involve special observational methods or equipment, and, if a large number of operators are available, a sizeable batch of statistics can be accumulated in short order.

An important limitation of the technique is that we cannot place much confidence in the relative magnitudes of various kinds of difficulties which come out of the study. For example, we cannot assume that frequencies such as are shown in Table 6 are representative of the frequencies with which these difficulties appear in the normal course of flying. Studies cited by Flanagan[53] show that there is a selective recall of critical incidents as a function of time (that is, some kinds of incidents tend to be soon forgotten), that the total number of incidents recalled is a function of time, and that the kinds of incidents recalled depend to some extent on who does the reporting. Finally, we can make no statements about the relative frequencies of various errors unless we know something about the exposure probabilities. There may be fewer opportunities for certain kinds of errors to occur.

SOME DIFFICULTIES WITH THE
STUDY OF ACCIDENTS

Like the methods of direct observation, the study of accidents and near accidents helps us to locate many sources of trouble in man-machine systems. As a method of study, however, it also has some serious limitations. Many of these have been mentioned above. The following are some general limitations to keep in mind:

1. Accidents, or near accidents, are infrequent events and it may take years to collect enough instances to form any sort of reasonable classification of the kinds of troubles that occur.

2. It is virtually impossible to get any meaningful comparisons of the absolute or relative frequencies of occurrence of various kinds of accidents because of the difficulty of estimating *exposure to the risk of accidents*. To take a fictitious example, suppose you are told that during 1957 there were 100 deaths from falls off stepladders in a certain city and 1,000 deaths from falls in bathtubs. Does this mean that bathtubs are *relatively* more dangerous than stepladders? Not necessarily. If the people of that city took a total of 1,000,000 baths during the year, the relative frequency of accidents of this sort is 1,000/1,000,000, or 0.1 per cent. But if stepladders were used only 1,000 times during the year, the relative frequency of accidents due to this source is 10 per cent. In both instances, the denominators of these ratios are critical but, as you can readily see, impossible to obtain.

Even these simple computations do not tell the whole story. Let us suppose that each bath lasted only 15 minutes, whereas stepladders were used for 30 minutes. Now, you see, we should correct our relative frequencies for the fact that the person who uses a stepladder has, on the average, twice as long to have an accident as compared with the person who uses a bathtub. Here is still another complication. Suppose that everyone who takes a bath takes only two steps—one to get in, the other to get out. Suppose, on the other hand, that whenever someone uses a stepladder he takes an average of six steps. The person on the stepladder has three times as many chances to slip and fall as does the person in the tub. That is another factor that should enter into our calculations. Without introducing any further variables, you can perhaps see why it is so difficult to interpret accident statistics.

As an exercise, think about the following statistics. The National Safety Council reports that in 1956 a total of 4,900 women and 42,500 men were involved in fatal traffic accidents in the United States. Entirely aside from your prejudices in this matter, do these numbers mean that women are safer drivers than men? What are some of the things you would need to know before you can arrive at a meaningful interpretation of these numbers?

3. In some accidents it is difficult, or impossible, to reconstruct what happened. The circumstances leading up to many accidents which are presumed to have been due to *pilot-error* are matters of conjecture because the pilot is no longer with us to say what he did.

4. In the near-accident and critical-incident techniques, we must often rely on the memories of people who work with particular systems. Sometimes there are ways of checking to be sure that these things

really did happen; more frequently there are not. In addition, accidents and near accidents are emotion-arousing situations—the worst possible for doing objective observing. People also forget many of the little mistakes they make. In fact, there is evidence that even serious accidents can be easily "forgotten" or overlooked.

LeShan and Brame report some interesting case studies in this connection.[71] For example, during a study of accident proneness, one man revealed a half-dozen major accidents he had been involved in. Intensive interviewing failed to reveal any others. At the end of the interview, the examinee was asked to strip, and his body was examined for scars. A previously undisclosed scar on the right side of his chest was called to his attention. He then remembered that three years earlier a bulldozer had rolled over him injuring his back and breaking three ribs. Another man was leaving the interview when the interviewer noticed that the little finger on the right hand was bent. On further inquiry, the examinee "just remembered" that he had broken that finger the previous year.

LeShan and Brame feel that behavior of this sort is the rule rather than the exception. It is difficult to support this belief, of course, because it is impossible to know how many accidents a person has had and has forgotten. They report, however, that in one study subjects were asked to list all the accidents they had had. They were given plenty of time with a sympathetic listener. When the subjects were through, they were then exposed to further careful probing. Thirty out of 35 subjects recalled several more accidents under the additional intensive questioning.

5. The final point to note is that the study of accidents will not necessarily tell you about causes or the means for reducing the number of accidents. Some people have the idea that when you have completely described the circumstances surrounding a set of similar accidents, the cure will be obvious. Sometimes it is; more often it is not. The remedy is not always explicitly revealed by the data. For example, McFarland and Moseley undertook a detailed analysis of records of severe traffic accidents in the engineering and claim files of two large casualty insurance companies.[80] One of their conclusions is that:

> A large majority of the accidents occurred under the following conditions: at intersections, or on country highways, on a dry level road surface made of concrete or macadam, having two lanes and with no reported structural defects, in daylight hours, when traffic

was light. It should be noted that all of these conditions are favorable for safe operation.

The assignment of causes in human behavior is an extremely difficult thing to do. Take this example: A man has a protracted argument with his wife. He stamps out of the house to the nearest bar and drinks four highballs. He then decides to go for a ride. It is nighttime, there is a skim of snow on the ground, and the tires on our victim's car are smooth. In rounding a poorly banked curve at excessive speed, the right front tire blows out, the car leaves the road and is demolished. What was the cause of the accident? The argument? Drinking? Speed? The weather? The smooth tires? The blowout? The poorly designed highway? It is impossible to say, for if we had changed any one of these factors, perhaps the accident would not have happened. We have no way of assigning a "cause," even though we may have a complete description of the circumstances leading up to the accident. In fact, the chances are very good that a coroner, state policeman, minister, psychiatrist, and highway safety engineer would each find different causes in this event.

We may also remind ourselves of the data in Table 6. The fact that so many pilots report difficulty with multi-revolution dials does not tell us how we can reduce these errors. We merely know that these errors exist. Perhaps better training in dial-reading might have helped reduce these confusions. And it is not at all apparent from the data in Table 6 that a redesign of the dial will help. To decide this matter, an experiment is necessary. This is the crux of the matter: Accident data may give us hunches, or clues. To test these hunches, our safest bet is to try them out.

Chapter 4 | *Statistical methods*

The statistical methods supplement all of the research techniques discussed so far and are really indispensable to any human experimentation. The statistics used in Chapters 2 and 3 were so elementary that they required no special comment, but, as we shall see in this chapter and the ones which follow, the interrelationship between statistics and research design can become very intimate and complex. The purpose of this chapter is (a) to acquaint you with some of the computational methods we shall need later and (b) to show why you cannot, or should not, plan an experiment without also considering the way in which the data will be treated.

Statistical methods embrace an extremely broad range of techniques about which whole books are usually written. In this chapter we can barely scratch the surface of this subject, and at best we can hope to become familiar with only a few of the formulae useful for human engineering work. In reading about them, however, notice how they tie in closely with the way observations are collected. Chapter 5 will come back to this point time and time again.

STATISTICS HELP TO SUMMARIZE DATA

If you look closely at the literature in various scientific fields you will probably be impressed by the great use behavioral scientists make of statistics. There are several reasons for this. First and foremost is the fact that people differ—they differ in height, weight, reaction time, and in any of 1,001 different ways in which you might want to study them. Despite this variability, it is apparent that you can make some statements about people which are generally true. On the average, men are taller and heavier than women. In general, taller people weigh more than shorter ones. More men than women are color-blind. These are all general statements—statistical statements—which are true of people in the aggregate. Any one of them may be false if you pick three or four people at random from the population. But it is a sure bet that you will find each of these statements correct for random samples of reasonable size, for example, samples over a hundred.

Considerations like these form the basis for the approach typically used by the behavioral scientist. To get reliable statements in the face of human variability, he must depend on multiple observations collected on many people. This brings us to the first important use of statistics in research. Statistical formulae help us to summarize and to describe data. When the events we are concerned with are inherently variable, statistics help boil down the data so that we can see significant trends and relationships. In this descriptive usage, statistics in an almost literal sense help us to see the forest instead of the trees.

DISTRIBUTIONS IN TABULAR AND GRAPHIC FORM

The value of statistics for summarizing sets of data can be demonstrated by working through the following example. In talking about an experiment on stoves (page 9), I mentioned that experimental subjects were required to push one of four controls on the front of the stove as quickly as possible after receiving a signal. Each subject made 80 such consecutive responses. Table 7 shows the 80 measurements recorded for one subject on Stove 2. By carefully scanning through the entries you can see that the longest measurement was 1.80 seconds and the shortest 0.60 seconds. Aside from this, however, it is hard to see any general trend in the measurements.

Frequency distributions. The first thing a statistician usually does with data like these is to arrange and group them numerically in a way which is more easily grasped. He compresses the data into classes and constructs a frequency distribution. The actual mechanics are illustrated in Figure 18. The range of values—0.60 to 1.80 seconds—is divided into about 12 to 18 classes. Fewer than about 12 classes generally means too great a compression of the data with some loss of precision in later computations; more than about 18 classes generally means too much detail. In this example, the data lend themselves well to 13 classes.

In this experiment the reaction times were read off an electric stop clock which was graduated in hundredths of a second. When the pointer on the clock stopped between two graduations, the reading was rounded off to the nearest hundredth. This means that a reading of 0.60 sec-

TABLE 7. *Reaction times (in seconds) made by one subject in responding to the controls of stove 2 (Figure 2)*

1.34	1.10	0.82	0.96
0.76	1.21	1.60	1.03
0.67	1.41	0.90	1.00
0.70	0.82	1.22	0.85
1.14	0.85	0.94	0.95
1.16	0.92	0.77	0.94
0.78	0.89	0.91	0.77
0.85	1.28	1.10	1.08
1.16	0.95	0.98	0.88
1.24	0.97	0.99	0.75
1.35	1.15	0.85	1.05
0.87	0.95	0.85	0.84
0.90	0.93	0.98	0.92
0.95	0.88	0.89	0.72
1.17	0.80	1.38	0.87
0.99	0.60	0.95	0.82
1.23	0.82	1.07	0.75
0.86	1.00	0.88	0.83
0.85	1.14	0.91	0.87
1.80	1.21	1.15	0.96

onds, for example, really represents times which range from 0.595 to 0.605 seconds. The class limits in Figure 18 are written in accordance with this meaning. As a rule statisticians do not write out the class limits as meticulously as they are shown in Figure 18. For simplicity these limits would normally be abbreviated to 0.60–0.69, 0.70–0.79, 0.80–0.89, etc. No matter how the class limits happen to have been expressed in a frequency distribution, you must use their exact values when you do certain calculations on the data (such as shown in Table 9 and Formula [4], for example).

After the class limits are selected and arranged, the raw data (from Table 7 in this example) are entered into the frequency distribution by tallying each item in the appropriate class as illustrated in the figure. Notice that groups of four tallies are crossed through with the fifth

Class Limits (Seconds)	Tallies	Frequencies
1.795 - 1.895	Y	1
1.695 - 1.795		
1.595 - 1.695	/	1
1.495 - 1.595		
1.395 - 1.495	Y	1
1.295 - 1.395	̷̷/̷/	3
1.195 - 1.295	̷̷̷̷/̷/1	6
1.095 - 1.195	̷̷̷̷/̷ IIII	9
0.995 - 1.095	̷̷̷̷/ I	6
0.895 - 0.995	̷̷̷̷/ ̷̷̷̷/ ̷̷̷̷/ ̷̷̷̷/ I	21
0.795 - 0.895	̷̷̷̷/ ̷̷̷̷/̷̷̷̷/̷̷̷̷/ II	22
0.695 - 0.795	̷̷̷̷/ III	8
0.595 - 0.695	̷/̷/	2
Total		80

FIG. 18. Construction of a frequency distribution for the data in Table 7.

tally so that they can be counted more easily. The tallying is always done twice—the second time as a check on the first. Ticks are added to the tallies as each item is checked the second time through. In Figure 18 the first 40 items in Table 7 have been so checked.

By grouping data into a frequency distribution we both lose and gain something. We lose the ability to identify individual measurements within each class. In exchange for this loss, however, we have a much clearer idea of the way the measurements are grouped. We can now see readily, for example, that over half of the measurements lie between 0.795 and 0.995 seconds. We can also see that they tend to come to a sharp break at the lower end of the scale, but tend to tail off gradually at the higher values. This type of a skewed—nonsymmetrical— distribution is called *positively skewed*. It is typical of reaction-time data because there is a definite lower physiological limit to reaction

TABLE 8.   *Frequency distribution of the heights of 2,960 Air Force cadets (from Randall et al*[92])

Class limits (cm.)	Frequencies	Cumulative frequencies	Cumulative percentages
198–200.9*	1	2,960	100.0
195–197.9	2	2,959	100.0
192–194.9	7	2,957	99.9
189–191.9	42	2,950	99.7
186–188.9	87	2,908	98.2
183–185.9	226	2,821	95.3
180–182.9	358	2,595	87.7
177–179.9	522	2,237	75.6
174–176.9	575	1,715	57.9
171–173.9	467	1,140	38.5
168–170.9	377	673	22.7
165–167.9	208	296	10.0
162–164.9	68	88	3.0
159–161.9	17	20	0.7
156–158.9	3	3	0.1

* These measurements were actually made to the nearest millimeter.

times but no such upper limit. Nonsymmetrical distributions which tail off toward the lower values are called *negatively skewed*.

The advantage gained by compressing the data of Table 7 into the distribution in Figure 18 is not particularly impressive because there are only 80 observations involved. You can well imagine, however, that before the frequency distribution in Table 8 was constructed, it was almost literally impossible to make any sense of the 2,960 measurements which came out of that study.

Normal distributions. The distribution in Table 8 is very close to a perfect *normal distribution,* two idealized versions of which are graphed in Figure 21. Normal distributions are symmetrical and can be fitted adequately with the following equation:

$$y = \frac{N}{\sigma\sqrt{2\pi}} e^{-\frac{x^2}{2\sigma^2}}$$ [1]

where y = the height of the curve at any point x,

N = the total number of cases in the distribution,

σ = the standard deviation of the distribution,

x = the deviation of any measurement from the mean, that is, $x = X - M$, and

e = the base of the natural system of logarithms, equal approximately to 2.718.

Several of these values will be discussed at greater length below.

The normal distribution is important in statistical work for two reasons: (a) many biological and psychological measurements are distributed normally (or very nearly so); and (b) many statistics are normally distributed even when they are calculated from distributions of data which are skewed. For example, although reaction times (like those in Table 7) are typically skewed, *averages* of large samples of reaction times are normally distributed. The full implications of this are difficult to convey to the novice. Suffice it to say, the normal distribution lies at the heart of most probability statements of the type we shall meet later in this chapter.

Histograms and frequency polygons. Data in frequency distributions are often graphed to bring out their principal features. Two such graphing procedures are illustrated in Figure 19. The histogram is made by constructing a bar for each of the classes in the frequency distribution. Each bar is as wide as the class; its height is proportional to the number of observations in the class. The frequency polygon is made by connecting the midpoints of the bars in the histogram. Note that the polygon is brought down to zero at both ends of the graph.

Cumulative frequency distributions. This is another type of graph which is useful in human engineering work—particularly in dealing with body measurements. Table 8 shows how such a distribution is constructed. The frequencies in the second column are cumulated in the third column. The cumulative percentages in the fourth column are obtained by dividing the cumulative frequencies by the total number of measurements, in this case, 2,960. The number 95.3, for example, means that 95.3 per cent of the men in this sample had heights of

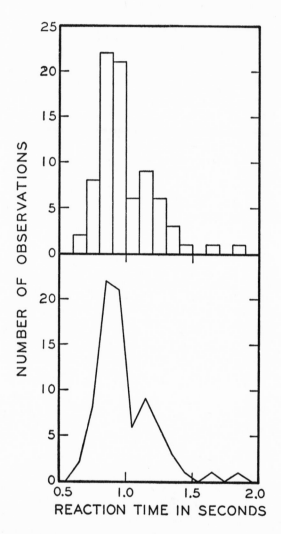

FIG. 19. A histogram (*above*) and frequency polygon (*below*) for the frequency distribution in Figure 18.

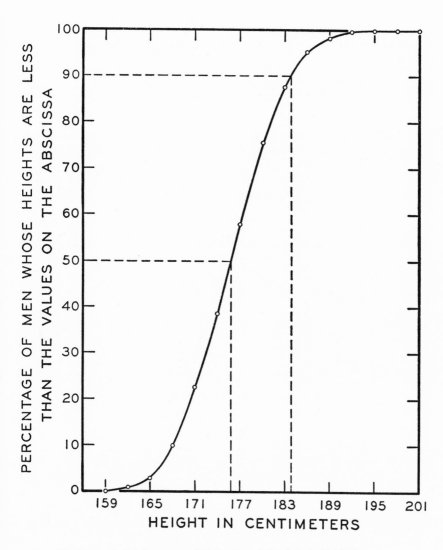

FIG. 20. A cumulative frequency distribution of the data in Table 8.

185.9 cm. or less. Conversely, only 4.7 per cent (100–95.3) of the men were taller than 185.9 cm. The other percentages in Table 8 are interpreted in the same way. Figure 20 shows the cumulative percentages in Table 8 plotted against height. Note that the cumulative percentages are plotted at the upper class limit in each instance.

MEASURES OF CENTRAL TENDENCY

Averages are numbers which can be used to stand for, or represent, a group of data. They generally fall near the middle of a distribution and so may be said to indicate the central tendency of the data. Statisticians use many averages in their work, but we shall define only the two most important ones here.

The arithmetic mean. The arithmetic mean, M, is what most people think of when they say average. Its formula is

$$M = \frac{\Sigma X}{N} \qquad [2]$$

where the Σ sign (the capital Greek sigma) means "the sum of"; X represents any measurement; and N is the number of measurements. To compute the average for the 80 measurements in Table 7 we merely add up all the numbers there and divide by 80. Since ΣX, the sum of the 80 reaction times, is 78.63, the arithmetic mean is 0.983 seconds.

When data are in a frequency distribution, statisticians make one assumption in computing the arithmetic mean. This assumption is that all of the cases in a class fall exactly at the midpoint of the class. The midpoints of the classes are shown in the second column of Table 9. To compute our ΣX now we assume that the two observations in the lowest class each have a value of 0.645, the eight observations in the second class each have a value of 0.745 and so on. Thus ΣX is equal to (2)(0.645) + (8)(0.745) + (22)(0.845) + (21)(0.945) + ⋯ + (1)(1.845), or 78.700. This sum divided by 80 gives us an arithmetic mean of 0.984. As you can see, the simplifying assumption had a negligible effect on the accuracy of this computation.

In actual practice, statisticians work with coded midpoints rather than true midpoints in order to reduce the labor of the computations. There are many ways in which one may code midpoints. Those shown in the last column of Table 9 are generally used for most routine computations. Note that they start with zero and go up as simple integers.

These coded midpoints were obtained by subtracting 0.645 from every true midpoint and then dividing the result by 0.1. In using coded midpoints we must remember to make suitable corrections to get the true mean value. $\Sigma X'$ in this problem is 271. This, divided by 80, gives us a value of 3.3875, which we may call the coded mean, M', to differen-

TABLE 9.   *Computing the mean and standard deviation for the data in Table 7*

Class limits (seconds)	Class midpoints	Frequencies (f)	Coded class midpoints
1.795–1.895	1.845	1	12
1.695–1.795	1.745		11
1.595–1.695	1.645	1	10
1.495–1.595	1.545		9
1.395–1.495	1.445	1	8
1.295–1.395	1.345	3	7
1.195–1.295	1.245	6	6
1.095–1.195	1.145	9	5
0.995–1.095	1.045	6	4
0.895–0.995	0.945	21	3
0.795–0.895	0.845	22	2
0.695–0.795	0.745	8	1
0.595–0.695	0.645	2	0

$$\Sigma X = 78.700 \qquad \Sigma X' = 271$$
$$\Sigma X^2 = 81.011 \qquad \Sigma(X')^2 = 1277$$

tiate it from the true mean, M. If we symbolize the calculation of the coded midpoints from the true midpoints in this way:

$$X' = \frac{X - b}{a} \qquad [3]$$

we must apply these transformations in the reverse order to go from the coded mean to the true mean. Thus:

$$M = (a)\,(M') + b \qquad [4]$$

or

$$M = (0.1)(3.3875) + 0.645 = 0.984.$$

As you can see, the use of coded midpoints gives us *exactly* the same outcome as we got working with true midpoints. Coded midpoints involve no hidden assumptions or approximations. They are merely a computational convenience.

As an exercise try computing the mean for the data in Table 8 using coded midpoints. Your answer should be 175.8.

The median. The median is another highly useful measure of central tendency that is commonly used when the data are markedly skewed. The median is that point in a distribution of measurements which divides the measurements exactly in half. Fifty per cent of the measurements are always above the median; fifty per cent below it. To compute the median for the data in Table 7, arrange all the reaction times in order from highest to lowest. Then count down to the fortieth measurement and take the point half way between it and the forty-first measurement as the median. This is equivalent to locating the value of the $\frac{N+1}{2}$ 'th measurement when they are arranged in order.

In this case, the computation is simple: The fortieth measurement, in order of size, is 0.94, and the forty-first measurement has the same value. Thus, the median is 0.94. Note incidentally that if there had been an odd number of measurements the median would be the value of the middle measurement when they are arranged in order. For the following numbers, 5, 6, 9, 3, 12, 1, 8, the median is the value of the fourth, $\left(\frac{7+1}{2}\right)$ 'th, one, or 6.

In computing a median for data in a frequency distribution, we use another assumption about the distribution of the cases within each class. This is that the observations are distributed uniformly throughout the class. To compute the median we find the value of the $\frac{N}{2}$ 'th observation by simple linear interpolation. The reason we look for the score of the $\frac{N}{2}$ 'th individual in this case, rather than that of the $\frac{N+1}{2}$ 'th individual, as in the example above, is that here we are working with grouped data in a *cumulative frequency* distribution. When we cumulate frequencies we add up all the frequencies below a certain point and that point is the upper limit of a score (refer to the discussion of this matter on page 98). When data are ungrouped, that is, not arranged in a frequency distribution, each number is usually the midpoint of a range of values and so must be increased by one-half to make it consistent with its corresponding cumulative value.

Table 8 shows that there are 1,140 individuals with heights of 173.9 cm. or less. The median is the measurement of the 1,480th one. Thus we must interpolate in the class, 174–176.9. A formula for doing this is the following:

$$Med = L + \frac{(\frac{1}{2} N - F) \, i}{f} \qquad [5]$$

where

L = the lower real limit of the class containing the median,
N = the number of items,
F = the total number of items below the class containing the median,
i = the size of the class interval, and
f = the number of items in the class containing the median.

As applied to Table 8 the computations are:

$$Med = 173.95 + \frac{(1,480 - 1,140)(3)}{575} = 175.7 \text{ cm.}$$

As a check, compute the median from above. The computations are:

$$176.95 - [(1,480 - 1,245)/575] \, [3] = 175.7.$$

As an exercise, compute the median for the data in Table 9. Do your computations from below and check them by working from above. Your answer should be 0.933.

The use of the mean and median. Although the mean and median are both highly useful statistics there are occasions when one is more appropriate than the other. In skewed distributions, for example, the mean tends to be more heavily influenced by extreme measurements than does the median. Thus, in a positively skewed distribution the mean is greater than the median, as was true for the reaction time data in Table 7. Because of these relationships the difference between the mean and median forms the basis for one simple test of skewness. When the data are so skewed that there is a substantial difference between the mean and median, the median is generally the fairer statistic to use if you want to show where the bulk of the data lie.

The mean and median are identical, or very nearly so, for normally-distributed data such as appear in Table 8. Under these circumstances either measure will represent the central tendency in the data fairly. The median is generally easier to compute, but it cannot be manipulated in algebraic equations as the mean can be (see, for example, equation [14]).

Having computed a measure of central tendency, the statistician usually wants to have some measure of dispersion to tell him how closely the data are grouped around the measure of central tendency. It is possible to have two distributions with identical means and medians but with different amounts of dispersion. The data in the two distributions in Figure 21 have the same means and the same numbers of observations. But the data in the tall slender distribution are much more closely grouped around the mean than are the data in the other distribution. As with measures of central tendency, there are a number of different measures of dispersion available to the statistician. We shall consider three here.

The range. The simplest measure of variation is the range, the difference between the smallest and largest observations in a series. The range is easy to compute and it answers a question we often intuitively ask about a set of data. It is somewhat less stable than other measures of variability, because it is computed from only two cases in a distribution and so can vary widely due to chance variations in the data. The range is 1.20 seconds for the data in Table 7 and approximately 42 cm. for the data in Table 8.

The standard deviation. This is by far the most useful measure of variability for general purposes. Its formula is:

$$\sigma = \sqrt{\frac{\Sigma x^2}{N}} \qquad [6]$$

where x is the deviation of any score from the mean of the distribution, that is, $x = X - M$. To apply it to the data in Table 7 we would need to compute $(1.34 - 0.983)^2 + (0.76 - 0.983)^2 + (0.67 - 0.983)^2 + \cdots + (0.96 - 0.983)^2$, divide this sum by 80, and then extract the square root of the quotient. In actual practice this is much too tedious a procedure because it involves 80 separate subtractions and 80 numbers to be squared. Statisticians usually use one of the two following formulae for the standard deviation:

$$\sigma = \frac{1}{\sqrt{N}} \sqrt{\Sigma X^2 - \frac{(\Sigma X)^2}{N}} \qquad [7]$$

$$\sigma = \frac{1}{N} \sqrt{N \Sigma X^2 - (\Sigma X)^2} \qquad [8]$$

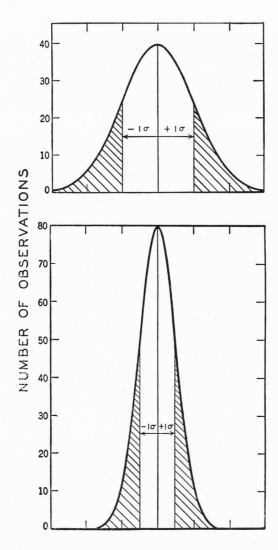

FIG. 21. Two distributions with identical means but with different standard deviations.

Formulae [6], [7], and [8] are all exactly equivalent and will give identical results. In applying Formula [8] to the data of Table 7 we need to get ΣX and ΣX^2. Both of these quantities can be obtained at the same time in a couple of minutes with a desk calculator. It is this convenience which makes formula [8] so useful in actual practice.

To compute ,the standard deviation for the data in Table 7 we find $\Sigma X = 78.63$ and $\Sigma X^2 = 80.6847$. The rest of the computations proceed as follows:

$$\sigma = \frac{1}{80} \sqrt{(80)(80.6847) - (78.63)^2}$$

$$= \frac{1}{80} \sqrt{272.0991} = \frac{1}{80} (16.495) = 0.206$$

The use of the standard deviation. At this point we can begin to get some appreciation for the summarizing function of statistical measures. Just two numbers, the mean and standard deviation, usually do a pretty good job of summarizing an entire group of data. For example, by consulting the appropriate statistical tables (tables of the normal probability integral), we will find that the mean plus-and-minus one standard deviation include about 68.3 per cent of the cases in a normal distribution. As applied to this example, 68.3 per cent of 80 cases equals 55 cases. The mean plus one standard deviation is 1.189 (that is, $0.983 + 0.206$). The mean minus one standard deviation is 0.777 (that is, $0.983 - 0.206$). Thus, we should expect to find about 55 cases between the values of 0.78 and 1.19. By actual count there are 59 cases. Incidentally, the area covered by plus-and-minus one standard deviation is illustrated in Figure 21.

To continue, in a normal distribution the mean plus-and-minus two standard deviations usually includes 95.4 per cent of the observations, or, translated into this example, we should expect to find 76 cases between the values of 0.57 and 1.40. By actual count there are 77 cases. The mean plus-and-minus three standard deviations includes about 99.7 per cent of cases or, translated into this example, we should expect to find all 80 observations between the values of 0.36 and 1.60. The actual count is 79 cases.

Similar predictions can be made for any other percentages. For example, tables of the normal probability integral show that $M \pm 0.6745\sigma$ should include the middle 50 per cent of cases. That is, we should expect to find 40 cases between the values 0.84 and 1.12. Actual count shows that there are 45 cases.

These predictions are not as good as is usually the case, because we are working here with a skewed distribution. Even at that, you will probably agree that these two statistics serve a very useful function in summarizing this set of data. When the distribution of measurements is normal in form and when there is a large number of observations, surprisingly accurate predictions about the observations can be made from just the mean and standard deviation.

Computing the standard deviation from a frequency distribution. The economy afforded by the use of coded midpoints becomes very evident when you compute the standard deviation for data in a frequency distribution (as in Table 9). Using the true midpoints, ΣX^2 is equal to $(2)(0.645)^2 + (8)(0.745)^2 + (22)(0.845)^2 + \cdots + (1)(1.845)^2$ or 81.011. The computation of the standard deviation proceeds as follows:

$$\sigma = \frac{1}{80} \sqrt{(80)(81.011) - (78.7)^2}$$

$$= \frac{1}{80} \sqrt{287.19} = \frac{1}{80} (16.947) = 0.212$$

Using coded midpoints, the calculation of $\Sigma(X')^2$ is greatly simplified, because it is simply $(2)(0)^2 + (8)(1)^2 + (22)(2)^2 + \cdots + (1)(12)^2$ or 1,277. The coded standard deviation is calculated as follows:

$$\sigma' = \frac{1}{80} \sqrt{(80)(1,277) - (271)^2}$$

$$= \frac{1}{80} \sqrt{28,719} = \frac{1}{80} (169.47) = 2.12$$

To get from the coded standard deviation to the true standard deviation we need to multiply only by the a of equation [3]. If you think about it you will see that adding or subtracting the same constant to every measurement in a set of data does not affect their variability. This is why we can ignore the b of equation [3] in going from the coded to the true midpoint. Thus:

$$\sigma = (0.1)(\sigma') = (0.1)(2.12) = 0.212$$

Once more note that the use of coded midpoints gives us exactly the same result as we obtained with the true midpoints.

In making these computations we again made use of the assumption that all of the observations in a class are exactly at the midpoint of the class. This means some small loss in accuracy, because the standard deviation computed from the frequency distribution in Table 9 is 0.212, whereas the standard deviation computed for the ungrouped measurements is 0.206. For all practical purposes, however, this is a trivial discrepancy since it amounts only to 6 in the third decimal place.

Sum-of-squares and variance. Before leaving the standard deviation, note that the quantity Σx^2 in formula [6] is known as the *sum-of-squares* and that it will be used time and again in later calculations. By comparing formulae [6] and [7] you will also see that:

$$\Sigma x^2 = \Sigma X^2 - \frac{(\Sigma X)^2}{N} \qquad [9]$$

The equation on the right of the equal sign is the one most commonly used for computing the sum-of-squares.

A variance is simply a standard deviation squared, that is, σ^2. A useful property of variances is that they can be added or subtracted in complex sets of data. This property will be exploited later in this chapter.

Percentiles. Percentile measures are commonly used in human engineering work to give some indication of the variability of measurements—particularly measurements of various body sizes. For example, a typical specification might state that a piece of equipment should be constructed so that it will accommodate all but the largest 10 per cent of a certain population. To turn this statement around, it means that the equipment should accommodate the smallest 90 per cent. The number we need is the 90th percentile.

Percentiles are computed in exactly the same way as medians are. The median is, in fact, a percentile itself—the 50th percentile. To compute the 90th percentile for the heights in Table 8, we must locate the height of the 2,664th individual (0.90 × 2,960). It is located in the class, 183 − 185.9, and is 69/226'ths of the way above the lower class limit.

$$90\text{th percentile} = 182.95 + \frac{(2664 - 2595)(3)}{226}$$

$$= 182.95 + \frac{69}{226}(3) = 183.9 \text{ cm.}$$

Note, incidentally, that percentiles can be obtained graphically with fair precision simply by interpolation in the cumulative frequency curve. The median and 90th percentile are illustrated in Figure 20. Other percentiles are, of course, computed on the analogy of those above.

The discussion so far has been concerned with statistics which describe measurements that vary along a single scale—for example, time, error in yards, or height. Frequently, however, the investigator may be interested in describing the *relationship* between measurements of one type and those of another. Height and weight are a good example. By looking at people you can tell that there is a relationship between height and weight: As a general rule, taller people weigh more than shorter people. You can also tell that the relationship is not perfect. Some people who are 5 feet tall weigh more than others who are 6 feet tall. How good is the relationship between height and weight? That is what correlations measure—the closeness of the relationship between measurements of one kind with those of another.

Speed and accuracy in operating a radar. A number of years ago I performed a series of experiments on various methods of operating a *VJ* remote radar indicator. The principal experimental conditions and some of the findings are shown in Table 10. It is not essential for our purposes at the moment to explain what these various operating conditions were, but you can see from the table that tests were made with single and multiple targets, with and without range rings, and at three different antenna rotation rates. The same equipment was used throughout the entire series of experiments. The same radar operators were used in every experiment, but they were tested according to a carefully prearranged plan which guarded against long-term changes in performance caused by such factors as learning and boredom. (Techniques for arranging such experimental schedules are discussed later in Chapter 5.)

The data we are concerned with right now are in the last two columns of Table 10. In the column headed X are the average times required by the radar operators to get bearing (direction) and range (distance) information for a single target. In the column headed Y are the average errors made by the radar operators in the ranges of the targets. So, for example, in experiment 8, it took the radar operators 11.2 seconds, on the average, to extract bearing and range information about a single

TABLE 10. *Summary of ten experiments made with the VJ Radar under various operating conditions (from Chapanis*[23])

Experiment number	Total number of targets tested	Experimental conditions	Average time per target (in seconds) (X)	Average range error (in yards) (Y)
1	90	Single targets; range rings visible	12.6	145
2	90	Single targets; no range rings	12.7	169
3	540	Multiple targets; range rings visible	9.9	198
4	537	Multiple targets; no range rings	10.2	204
5	60	Single targets; 4 RPM	14.2	133
6	60	Single targets; 6 RPM	12.9	163
7	60	Single targets; 12 RPM	10.8	175
8	360	Multiple targets; 4 RPM	11.2	182
9	360	Multiple targets; 6 RPM	9.6	195
10	357	Multiple targets; 12 RPM	9.4	227

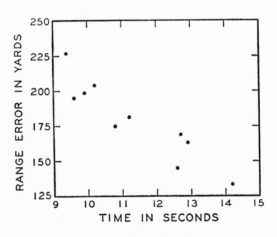

FIG. 22. A correlation diagram of the data in Table 10.

target appearing on the radar scope. The ranges reported by the operators had an average error of 182 yards. These averages are based on a total of 360 targets studied in that particular experiment.

A plot of these data, in Figure 22, shows that there is a negative relationship between time and errors: Those operating conditions which gave the longer average times yielded the smaller errors; those conditions which gave the shorter times were associated with bigger errors.

The Pearson product-moment coefficient of correlation. A measure used widely for data like these is the Pearson product-moment correlation. Three mathematically equivalent variations of its formula are:

$$r_{XY} = \frac{\Sigma xy}{N \sigma_X \sigma_Y} \qquad [10]$$

$$r_{XY} = \frac{\Sigma(X - M_X)(Y - M_Y)}{N \sigma_X \sigma_Y} \qquad [11]$$

$$r_{XY} = \frac{N\Sigma XY - (\Sigma X)(\Sigma Y)}{\sqrt{N\Sigma X^2 - (\Sigma X)^2} \sqrt{N\Sigma Y^2 - (\Sigma Y)^2}} \qquad [12]$$

When there are only a few data to be computed, and when there is a standard desk calculator available, it is generally easiest to use formula [12]. For the data in Table 10, verify the following calculations for yourself:

$$N = 10 \qquad \Sigma XY = 19,939.8$$

$$\Sigma X = 113.5 \qquad \Sigma Y = 1,791$$

$$\Sigma X^2 = 1,312.75 \qquad \Sigma Y^2 = 327,967$$

$$r = \frac{10(19,939.8) - (113.5)(1,791)}{\sqrt{10(1,312.75) - (113.5)^2} \sqrt{10(327,967) - (1,791)^2}}$$

$$= \frac{-3,880.5}{\sqrt{245.25} \sqrt{71,989}} = \frac{-3,880.5}{(15.660)(268.31)} = \frac{-3,880.5}{4,201.7}$$

$$= -0.924$$

Interpreting the correlation coefficient. Correlation coefficients range from +1.0 through 0 to −1.0. Positive and negative coefficients of the same numerical value are equally good. An r of −0.8 indicates as much correlation as one of +0.8; the sign merely tells us the direction of the relationship (see Figure 23). A zero correlation means that there is no relationship between the two sets of scores, and the higher the absolute value of the coefficient the closer the relationship. An r of 0.8 indicates a closer relationship than one of 0.4, but the coefficients do not tell

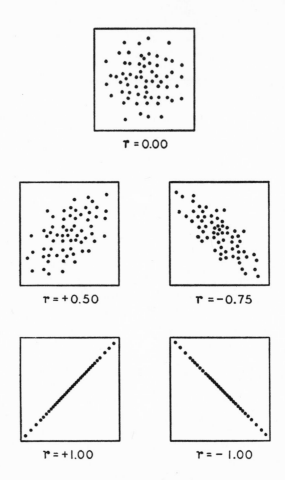

FIG. 23. This figure illustrates correlation coefficients of various amounts.

you the amount of relationship directly. They are not percentages, and an r of 0.8 is not twice as good as one of 0.4. One way of expressing the "goodness" of a correlation is to show how much a given correlation increases our predictability, or conversely, how much it reduces errors in our predictions.

Figure 24 is a schematic diagram of a correlation between X and Y. The ellipse is an idealized contour of the type which is usually found to enclose all the data in a correlation diagram (such as in Figure 22) when a very large sample of data is plotted in the diagram. Distribution A along the right-hand edge of this figure is the normal distribution we would get if we were to plot all the Y's without regard to their

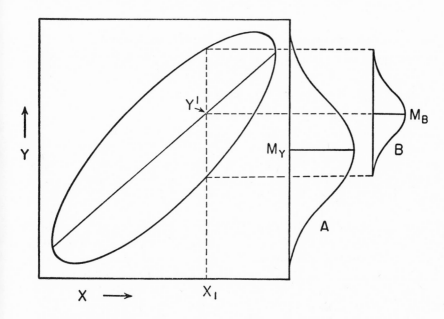

FIG. 24. This diagram illustrates how a correlation reduces the possible error in predicting Y if X is known.

X values. If you came up to a statistician and said, "I have in mind a certain item in that set of data. What is your best guess about its Y-value?" the statistician's best estimate would be that it has a value equal to M_Y. But it *could* have a value anywhere within that distribution, and the size of the distribution is a measure of the possible error the statistician might make if he knew nothing more about the specimen. As we have already seen, there is a statistical measure, σ_Y, we can use to quantify the amount of that error.

But suppose you had measured the X-value of the specimen and found that it had a value of X_1. Now what is the statistician's best estimate of its Y-value? If the statistician knows the magnitude and form of the correlation, he can reduce his error of estimate considerably. If a specimen has a value X_1, the best estimate of its Y-value is Y', the mean of that segment of the vertical dashed line which falls within the correlation ellipse. Y' is also the mean of the smaller distribution B, shown at the far right. Distribution B, of course, is the distribution of all those Y's that have an X-value of X_1. The error of predicting Y when you know X is σ_B. The average of all the σ_B's for all such possible small distributions in the correlation diagram is $\sigma_{est Y}$ and is called the *standard error of estimate.*

The ratio of $\dfrac{\sigma^2_{est Y}}{\sigma^2_Y}$ is the proportion of the total variance in Y-scores which cannot be predicted from X scores, and $1 - \dfrac{\sigma^2_{est Y}}{\sigma^2_Y}$ is the proportion of the total variance in Y scores which *is* predictable from X scores. The latter value turns out to be equal to r^2, that is,

$$r^2 = 1 - \frac{\sigma^2_{est Y}}{\sigma^2_Y} \qquad [13]$$

or in words r^2 represents the proportion of the total variation in Y which is predictable from X. For this reason r^2, rather than r, is said to represent the amount of the relationship between the two variables. A correlation of 0.707 ($r^2 = 0.50$) is thus twice as good as one of 0.50 ($r^2 = 0.25$), because the errors of estimate in the former case are only half what they are in the latter. When $r = +1.0$ or -1.0, the data fall exactly on a straight line and there are no errors of prediction. When $r = 0.0$, the errors of prediction are maximum and $\sigma_{est Y} = \sigma_Y$.

The diagonal line in Figure 24 is the *regression line* for predicting Y from X. It is also called the best-fitting straight line for the ellipse, because the standard error of estimate, $\sigma_{est Y}$, computed around this line is smaller than the standard error of estimate computed around any other straight line you might try to fit to the data. Its equation is:

$$Y' = r\frac{\sigma_Y}{\sigma_X} X + \left(M_Y - r\frac{\sigma_Y}{\sigma_X} M_X\right) \qquad [14]$$

where Y' is the Y predicted for any X, and all the other terms have the meanings previously defined in this chapter. There is another regression line for predicting X from Y. Its equation is obtained by interchanging X and Y everywhere in equation [14].

A measure of partial correlation. Some statistical measures come very close to being experimental techniques in the sense that they enable the experimenter to untangle the effects of numerous variables in complex sets of data. Such a technique is the partial correlation. A partial correlation tells you the amount of relationship between two variables when the effects of one or more other variables are held constant. For example, $r_{12.3}$* is a *first-order* partial correlation. It indicates the amount of relationship between variables 1 and 2 when the influence of a third variable, 3, is held constant. Another way of saying this is that $r_{12.3}$ is the correlation you would expect to find between 1 and 2 if you had picked only those 1's and 2's which had the same value on variable 3.

To continue, $r_{12.34}$, a *second-order* partial correlation, indicates the amount of relationship between variables 1 and 2 when the effects of 3 and 4 are held constant. Higher-order partial correlations are written on the pattern of those two above. A simple, two-variable correlation, r_{12}, is also called a *zero-order* correlation. It makes no difference in what order you write subscripts on either side of the period. Interchanging variables across the period, however, changes the meaning of the correlation. That is,

$$r_{12.34} = r_{12.43} = r_{21.43}$$

but

$$r_{12.34} \neq r_{13.24}$$

and

$$r_{12.34} \neq r_{34.12}$$

The formulae for two first-order partial correlations are as follows:

$$r_{12.3} = \frac{r_{12} - r_{13}\, r_{23}}{\sqrt{(1 - r^2_{13})(1 - r^2_{23})}} \qquad [15]$$

$$r_{13.2} = \frac{r_{13} - r_{12}\, r_{23}}{\sqrt{(1 - r^2_{12})(1 - r^2_{23})}} \qquad [16]$$

With these as models, you should be able to write the equation for the third one, $r_{23.1}$.

A general formula for partial r's of any order is the following:

$$r_{12.34\,\cdots\,n} = \frac{r_{12.34\,\cdots\,(n-1)} - r_{1n.34\,\cdots\,(n-1)}\, r_{2n.34\,\cdots\,(n-1)}}{\sqrt{(1 - r^2_{1n.34\,\cdots\,(n-1)})(1 - r^2_{2n.34\,\cdots\,(n-1)})}} \qquad [17]$$

Partial correlation in a study of highway accidents. The partial correlation method is particularly suited to situations where the inves-

* I have switched from the alphabetic subscripts used in formulae [10], [11], and [12] only because numerical subscripts are handier to use in formulae [15], [16], and [17] which follow. In any case a letter or numeral refers to a single variable.

tigator cannot arrange conditions of observation to suit himself but must take the data as they come. Lauer and McMonagle, for example, report a study made by the Michigan State Highway Department on accidents along a trunkline highway and their relation to design and roadside features.[70] One objective of the study was to discover whether road signs affect accident rates. It was impossible to do a carefully controlled experiment on this problem, because the investigators had no control over the density of the traffic or of the advertising signs along selected stretches of the highway. What they did was to select a 100-mile stretch of highway for test, make a comprehensive survey of all the design and roadside features in which they were interested, and keep a meticulous record of all accidents that occurred along the test highway. (During a two-year period there was a total of 1,968 accidents reported along the test highway.)

The 100-mile stretch was divided into 119 sections each of which contained intersections with secondary roads. The accident rate and the sign density were then computed for each section. When these two measures were correlated for the 119 sections, the coefficient of correlation amounted to $+0.712$. This result seems to indicate a high positive relationship between the density of signs and accident rates.

The picture is not as simple as that outcome suggests, however. The investigators had to take into account the fact that the density of taverns, gas stations, commercial garages, stores, restaurants, and other establishments also varied in the 119 test sections of the highway. In addition, there were variations in the number of private driveways and other regular highway design features (transitions in the width or arrangement of lanes, culvert posts, elevations, and the like) to cloud the interpretation. What the authors really wanted was the correlation between the density of advertising signs and accident rate, with all those other extraneous variables held constant. This partial correlation turned out to be only 0.002*—or no correlation at all. By contrast, the zero-order correlations between accident rate and density of taverns ($+0.698$) and between accident rate and density of gas stations and commercial garages ($+0.666$) were both less than that between accident rate and sign density. The partial-correlation analysis showed, however, that these two factors—the density of taverns and of gas stations—were most highly correlated with accident rates when all other factors were held constant.

---

* Since the report from which these results are quoted does not give the raw data in sufficient detail to illustrate the arithmetic of these computations, we must be content with only a general summary of the findings.

THE EVALUATION OF
EXPERIMENTAL OUTCOMES

The foregoing material in this chapter has been concerned with the descriptive, summarizing, or condensing function of statistical measures. We turn now to the second important use of statistics, namely, in helping us to evaluate the significance of the differences or effects we find in our research. There are basically two kinds of questions we should be asking when we ask about the significance of results. The first is, "What is the statistical significance of these results?" The second is, "What is the practical significance of these results?"

EVALUATING THE STATISTICAL SIGNIFICANCE OF
EFFECTS: THE t-TEST

We spoke about the problem of human variability earlier in connection with the summarizing function of statistics. But there is another consequence of human variability which is even more important. In human research you can be certain that no two experimental outcomes will be exactly the same. Even if you repeat an experiment under exactly the same conditions, you will not get the same results the second time as you did the first. Sometimes the difference you find between two experimental outcomes is simply the result of normal variability, that is, it is a chance result—the result of random fluctuations. Sometimes, however, the difference in outcomes is the result of underlying factors which produced a real difference. The problem of evaluating the statistical significance of effects, then, is essentially the problem of distinguishing between those outcomes which are the results of chance variability and those which are genuine.

What we mean when we say that we want to know about the statistical significance of a difference is this: Could random variability, or chance, account for an effect as large as this? This is a very sensible question, because there is no point worrying about an effect if it could be explained by the random variability inherent in the data.

The basic ingredients of a statistical test. There are essentially four basic quantities that enter into a test of statistical significance. The first is the size of the difference. Other things equal, the bigger the difference the more confidence we have in its genuineness. The second is the number of observations contributing to the effect. Given a difference of a certain size, we have more confidence in it if it is based on 100 people rather than on 10 people. The third factor is the amount of random variability in the measurements. The less the variability, the greater the confidence we have that a difference of a certain size is genuine. Combining the foregoing three quantities in appropriate formulae yields the fourth measure—a number from which we can get some indication of the probability, or betting odds, that a difference of this size, based on so many cases, and with so much variability could have occurred through the normal workings of the laws of chance. Formulae for testing the statistical significance of effects are essentially ways of combining the first three factors into a single index which can then be evaluated by consulting the appropriate probability tables.

To put this discussion into more exact language, consider the following set of formulae. A common problem in experimentation is that of evaluating the statistical significance of the difference between two means. If the samples are large (100 cases or more) and made up of two independent groups of subjects, the evaluation is done by calculating the number z, from

$$z = \frac{M_1 - M_2}{SE_{Diff(M_1-M_2)}} \qquad [18]$$

where M_1 and M_2 are the two means, and $SE_{Diff(M_1-M_2)}$ is the standard error of the difference between the means. The latter can be found from the following equation:

$$SE_{Diff(M_1-M_2)} = \sqrt{\frac{\sigma^2_1}{N_1} + \frac{\sigma^2_2}{N_2}} \qquad [19]$$

where σ_1 and σ_2 are the standard deviations of the two sets of data and N_1 and N_2 are the numbers of observations in the two sets of data.

Combining all these into a single equation gives us:

$$z = \frac{M_1 - M_2}{\sqrt{\frac{\sigma_1^2}{N_1} + \frac{\sigma_2^2}{N_2}}} \qquad [20]$$

This formula describes more precisely the relationships discussed in the preceding section. The z itself is evaluated by consulting tables of the normal probability integral. The larger the value of z, the less likely it is that the observed difference between M_1 and M_2 could have occurred by chance.

Comparison of two independent means. When *small* samples are being compared the appropriate formulae are a little different from [18], [19], and [20], although the basic idea is the same. I shall illustrate the procedure with an example.

In a study of a synthetic training device, the Link Trainer, Flexman, *et al.*, tested two groups of pilot trainees.[54] Both groups received exactly the same training curriculum except that one group was also given

TABLE 11. *Over-all flight-test grades assigned by a CAA examiner to two groups of pilot trainees (from Flexman, et al.*[54])

	Link group	Control group
	65	60
	70	65
	72	40
	70	76
	66	45
	80	70
	67	75
	68	52
	67	50
	70	45
	73	60
	69	65
	74	60
	71	60
	62	60
	60	50
	60	55
	76	60
	68	60
	60	70
	63	55
ΣX	1,431	1,233
ΣX^2	98,087	74,255
M	68.1	58.7

experience in the Link Trainer. After their training, both groups of subjects were given an independent flight examination by a Civil Aeronautics Administration examiner, who assigned grades to the student's flight performance. It is important to point out that the CAA examiner did not know what kind of training any of the students had received. Table 11 shows the flight test grades assigned to the two groups of students.

The computations at the bottom of the table show that the students who had received prior experience on the synthetic trainer received an average flight-test grade about 10 points higher than the average grade received by the other group. On the other hand, it is evident that there is a considerable amount of overlap in the scores earned by the two groups. The lowest grade received by any student in the Link group was 60. Thirteen of the 21 students in the other group received grades at least this high. So the question arises: "Is this difference between the two groups reliable?"

To make the appropriate statistical test we need first to compute the sum-of-squares for each of the two groups, using equation [9]. Verify for yourself that $\Sigma x_L{}^2 = 575$ and that $\Sigma x_C{}^2 = 1,860$. These two quantities are entered into the following formula for the standard error of the difference between the two means

$$SE_{M1-M2} = \sqrt{\left(\frac{\Sigma x_1{}^2 + \Sigma x_2{}^2}{N_1 + N_2 - 2}\right)\left(\frac{1}{N_1} + \frac{1}{N_2}\right)} \qquad [21]$$

The actual computations are:

$$SE_{M1-M2} = \sqrt{\left(\frac{575 + 1860}{21 + 21 - 2}\right)\left(\frac{1}{21} + \frac{1}{21}\right)}$$

$$= \sqrt{\frac{2435}{420}} = 2.41$$

The above result is then entered into this formula to obtain t

$$t = \frac{M_1 - M_2}{SE_{M1-M2}} \qquad [22]$$

As applied to this problem the computations are:

$$t = \frac{68.1 - 58.7}{2.41} = 3.90$$

On degrees of freedom. The t of 3.90 is then evaluated by consulting tables of the t distribution with $N_1 + N_2 - 2$, or 40, degrees of freedom. This concept of degrees of freedom is an important one which will keep coming up throughout the rest of the book. For all the problems in this book, the degrees of freedom, or df, refers to the number of x's in any Σx^2 which are free to vary.

Suppose that we were to calculate the Σx^2 for three scores whose M is 6. If M is 6 then, of course, the sum of the three scores, ΣX, must be

18. What possible values could the *first* score have? Obviously it is completely free to vary and could be anything at all. Let's assume that it is 35. What possible values could the *second* score have? It is also completely free to vary. Let's assume that it is 4. Now what possible values could the *third* score have? The third score is not free to vary because we were told at the beginning that ΣX equalled 18. This means that the third score can only be $(18 - X_1 - X_2)$ or -21. Since each x is a score minus the mean of the scores, only two of the x's are free to vary in this example.

In general, the df for any sum-of-squares is the number of squared terms minus the number of restrictions imposed by constants which are computed from the data. In applying formula [21] we used M_1 to compute Σx^2_1. That's one restriction. Then we used M_2 to compute Σx^2_2. That's the second restriction. Hence $df = N_1 + N_2 - 2$.

The evaluation of t. To return to our problem, tables of the *t*-distribution will show that a t of 3.90 is greater than the one for a probability or p of 0.001. What does this mean? Literally it means the following: If these two groups of data had been randomly selected from a single basic set (or population) of measurements, we would expect to find a difference between the two means as large as this less than once in 1,000. This is such a rare event that most people would be willing to reject the basic condition implied in the sentence above. The argument would go something like this: For me to have gotten an event which could only happen once in 1,000 times is too much for me to believe. The only way I can explain this low probability value is to say that my basic assumption is wrong, viz., these two groups of data were *not* drawn from the same basic set of measurements.

In actual practice, statisticians and experimenters do not go through this complex argument every time they evaluate an experimental outcome. In somewhat looser terms, they would probably say, "There's a real difference between these two groups. This difference is significant at less than the 0.1 per cent level."

On probabilities and levels of significance. The solution achieved above brings up a very important feature of statistical analyses. The statistician's answers always come out in terms of probabilities. It is up to the experimenter or engineer to decide what probability level he is willing to accept. If you were experimenting with the tolerable dose of a drug, you might not be satisfied unless the risk of a fatality were less than 1 in 10,000. If, on the other hand, you were concerned with the probability of getting defective pieces in a lot of manufactured items, you might be satisfied with much higher probabilities, for example, 1 in 10. The point is that these risks are things which the experimenter or consumer must set. For most experimental work, however, it seems to be generally agreed that differences are significant if they could occur by chance less than 5 per cent of the time (probability = 5 per cent, or 0.05). A difference is generally conceded to be *highly* significant if it could occur only once in 100 times by chance (probability = 1 per cent, or 0.01).

Let's look at another aspect of these probability notions a little more closely to see what they imply. Assume that we get an experimental result of a certain size and that it turns out to be significant at exactly the 5 per cent level. This means that a result as large as the one we found *could* have occurred by chance 1 in 20 times. In the long run, if we always set our criterion at the 5 per cent level of significance, we run the risk of saying that an experimental result is significant when it really is not. This will happen about once in every 20 experiments we run. In statistical terms, this is a Type I error. In the face of this the experimenter might say, "Well, let's not take chances like this. Let's insist that all of our differences must be significant at the 1 per cent level, or even the 0.1 per cent level, before we accept them as indications of true differences. In this way we will decrease the probability of saying that an experimental result is significant when it really is not."

This argument is sound, as far as it goes. The trouble with it is that there is another kind of risk the experimenter must consider. That is the risk of saying that an experimental outcome is not significant, when it really is. Statisticians call this a Type II error. Unfortunately, the chances of making Type I and Type II errors are inversely correlated. The lower the probability of making a Type I error, the greater the probability of making a Type II error, that is of saying that something is not significant when it really is. Elaborating on this point would take up many pages, but perhaps you can see the basic problem. It is essentially one of achieving a balance between these two kinds of errors. You want to set your probability levels high enough so that you do not make too many Type I errors; on the other hand, you do not want to set them so high that you start making an excessive number of Type II errors. For many practical experimental problems the 5 and 1 per cent levels of significance strike the right balance.

There are some practical situations in which the consequences of making one or the other type of error may be very serious. In these instances, the experimenter may want to select different probability levels. Suppose, for example, that you have an industrial plant equipped with a certain type of machine. Along comes someone with an alternative type of machine which is supposed to yield higher productivity. You plan to run an experiment to compare the old and new machines, and you expect to find that the new one is better. How certain do you want to be of this outcome? In this instance you would almost certainly want to set the probability level very low because the consequences of a wrong decision here are serious. A decision to take out all the old machines and install new ones represents such a large capital expenditure that you would want to be really sure that there was a difference between the two machines.*

On the other hand, there may be instances in which you might be willing to set the probability values higher than usual. If you are doing a series of exploratory studies on possible improvements in a process, you might want to set the probability values sufficiently high so that you will not overlook procedures or techniques which might be worth following up.

The final choice of probability values depends on the good judgment of the experimenter. In this matter, statistics can be of no help.

THE ANALYSIS OF VARIANCE

The F-test for comparing two means. The *t*-test described above is the procedure used most commonly for evaluating the significance of the difference between two means. There is another procedure, equally good, which can be used for the same purpose. This is the *F*-test based on an analysis of variance of the data. The sum-of-squares formula [9] forms the basis for this procedure. First, we compute the sum-of-squares for the entire group of data in Table 11. This is done in the following way:

$$\Sigma x^2{}_T = 98,087 + 74,255 - \frac{(1,431 + 1,233)^2}{42}$$

$$= 172,342 - 168,974 = 3,368$$

Note, incidentally, that the same term $\frac{(\Sigma X)^2}{N}$, often called the correction term, is used repeatedly throughout the calculations for any given problem.

* There are also some other practical considerations which would influence your decision, but for the moment, our interest is only in this basic statistical issue.

It is possible to break up this total sum-of-squares into two independent components. One part, the *between* groups sum-of-squares, measures the variation of the group means around the over-all mean. The other part, called the *within* groups sum-of-squares, is a measure of the variation of the scores within each group around their own group means. Thus the *within* groups sum-of-squares is a measure of the random variability in the measurements. When each of these sums-of-squares is divided by its appropriate number of degrees of freedom, we obtain two independent estimates of the same population variance. The ratio of these two estimates is the F-ratio. If there is no significant difference between the groups, we should have merely two independent estimates of random variability in the F-ratio so that, on the average, the F should be 1.00. On the other hand, if the group means really differ because of the influence of the experimental variable, there will be an extra source of variability in the numerator of the F-ratio so that, on the average, it should be considerably larger than 1.00. Tables of the F-distribution enable us to determine the probability that F values of various sizes could have occurred by chance.

To exemplify this procedure we compute a sum-of-squares for the differences between the two groups. This is done by using equation [9] again except that instead of the X we substitute the ΣX for each group. This gives us:

$$\Sigma x^2_{BG} = \frac{(1,431)^2}{21} + \frac{(1,233)^2}{21} - 168,974 = 169,907 - 168,974 = 933$$

The divisor of 21 comes from the fact that there are 21 items contributing to each of the numbers which gets squared in the numerator. This is a general rule which will hold throughout all the computations which follow. In the computation of the total sum-of-squares the divisor was not shown because it was 1. This same divisor is used in the two sums-of-squares computations immediately below.

Finally, we compute a sum-of-squares for the data within groups as follows:

$$\text{Within Link group} = 98,087 - \frac{(1,431)^2}{21} = 98,087 - 97,512 = 575$$

$$\text{Within control group} = 74,255 - \frac{(1,233)^2}{21}$$

$$= 74,255 - 72,395 = 1,860$$

$$\Sigma x^2_{WG} = 575 + 1,860 = 2,435$$

All these computations are entered into Table 12. Some important features of the entries in Table 12 are that the sum-of-squares term for the difference between groups and the sum-of-squares for the data within groups add up to the total sum-of-squares. This additivity of the sum-of-squares terms is an important characteristic of such analyses. Looked at the other way around, we have taken the total sum-of-squares and broken it down into two separate and independent parts.

A second important feature of Table 12 is that the degrees of freedom also add up. In an analysis of variance the number of degrees of freedom assigned to any source of variation is one less than the number of items contributing to the sum-of-squares term. In computing the

TABLE 12. *Analysis of variance for the data in Table 11*

Source of variation	Sum of squares	Degrees of freedom	Estimate of variance	F-ratio
Between groups	933	1	933	15.3
Within groups (error)	2,435	40	60.9	
Total	3,368	41		

total sum-of-squares there were 42 items. One less than this gives us 41 degrees of freedom. In computing the sum-of-squares for the difference between groups, we had two items: the sum for the Link group and the sum for the control group. Two minus one gives us one degree of freedom for this source of variance. In computing the degrees of freedom for the data within the Link group, we dealt with 21 items which give us 20 degrees of freedom. These, plus the 20 degrees of freedom from the other group, give us the 40 degrees of freedom we have assigned to the data within groups.

In Table 12, the estimated variance is computed by dividing the sum-of-squares by the appropriate degrees of freedom. The F-ratio is formed by dividing the variance between groups by the variance within groups. This ratio is 15.3. Consulting the appropriate probability tables under 1 and 40 degrees of freedom tells us that this difference is highly significant.

For the special case where there are only two groups involved, F turns out to be equal to t^2. Thus, for this special case, we can use either the F- or t-tests. They give us identical answers. When there are more than two groups being compared, we cannot use the t-test but must use the F-test.

Comparing three independent groups. The next logical step is to extend this statistical test to a comparison of three groups. An appropriate set

TABLE 13. *Reaction times (in seconds) earned by 45 subjects tested on the three stoves in Figure 2. Fifteen subjects were tested on Stove 1; 15 on Stove 2; and 15 on Stove 3 (unpublished data from Chapanis[28])*

	Stove 1	Stove 2	Stove 3
	0.52	0.67	0.41
	0.59	0.73	0.43
	0.35	0.56	0.84
	0.38	0.64	0.66
	1.35	0.92	0.43
	0.52	0.96	0.68
	0.60	0.60	0.50
	0.63	1.20	0.51
	0.68	0.81	0.68
	0.49	0.74	0.62
	0.54	0.93	1.05
	0.41	0.69	0.52
	0.46	0.72	0.98
	0.38	0.86	0.72
	0.57	0.73	0.82
ΣX	8.47	11.76	9.85
ΣX^2	5.5783	9.6086	7.0225

of data appear in Table 13. These were collected in the experiment on stoves mentioned in Chapter 1 (pp. 9 ff.). Forty-five different subjects participated over a series of 80 consecutive trials. The data in Table 13 are for the last trial only.

As you can see from the table, the average reaction time was fastest for stove 1 and slowest for stove 2, with stove 3 roughly between the two. Are the differences among these reaction times significant? The analysis proceeds exactly as the analysis for the Link Trainer experiment above. The computation for the sum-of-squares for the differences between stoves is again based on equation [9] except that the ΣX for each group is substituted for the X in the equation. So:

$$\Sigma x^2_{BS} = \frac{(8.47)^2}{15} + \frac{(11.76)^2}{15} + \frac{(9.85)^2}{15} - \frac{(8.47 + 11.76 + 9.85)^2}{45}$$

$$= 20.4707 - 20.1068 = 0.3639$$

Note again the general rule about the divisor 15: each of the sums in the numerator is based on 15 items. The total sum-of-squares is computed as follows:

$$\Sigma x^2_T = 5.5783 + 9.6086 + 7.0225 - 20.1068 = 2.1026$$

The sum-of-squares within stoves as follows:

$$\text{Within stove 1} = 5.5783 - \frac{(8.47)^2}{15} = 0.7956$$

$$\text{Within stove 2} = 9.6086 - \frac{(11.76)^2}{15} = 0.3888$$

$$\text{Within stove 3} = 7.0225 - \frac{(9.85)^2}{15} = 0.5543$$

$$\text{Within all three stoves} = 0.7956 + 0.3888 + 0.5543 = 1.7387$$

All these computations have been entered into Table 14. Note again that the sum-of-squares and the degrees of freedom for each of the component sources of variance add up to the total as before. The F-ratio shows that the differences between stoves are significant at the 5 per cent level.

With the examples in Tables 11, 12, 13, and 14 it should be clear how to generalize these analyses to experiments in which four or more experimental groups have been tested.

A subjects-by-experimental conditions analysis. In the stove experiment just analyzed, the 15 subjects who were tested on stove 1 were different from the 15 who were tested on stove 2 and the 15 who were tested on stove 3. The next level of complexity in design is an experiment in which the same subjects are tested under each of several different experimental conditions. Such a design is shown in Table 15. The data there came from an experiment on the effects of anoxia—oxygen lack at high altitudes—on vision. The visual test consisted of a series of Landolt rings (visual-acuity test objects) carefully graded in brightness contrast. The scores in Table 15 are the total number of

TABLE 14.   *Analysis of variance for the data in Table 13*

Source of variation	Sum of squares	Degrees of freedom	Estimate of variance	F-ratio
Between stoves	0.3639	2	0.1820	4.40
Within stoves (error)	1.7387	42	0.04140	
Total	2.1026	44		

TABLE 15. *Scores earned by 20 subjects on a test of visual contrast sensitivity under three conditions of anoxia (from Chapanis[19])*

Subject	2,000 ft., breathing air	15,000 ft., breathing air	15,000 ft., breathing oxygen	Totals
A	28	31	26	85
B	11	10	8	29
C	11	4	7	22
D	19	16	20	55
E	22	18	21	61
F	27	13	18	58
G	16	8	15	39
H	12	11	12	35
I	19	18	18	55
J	23	16	16	55
K	5	5	13	23
L	20	15	20	55
M	15	17	14	46
N	15	13	19	47
O	23	11	13	47
P	20	17	19	56
Q	19	13	19	51
R	15	11	15	41
S	22	16	16	54
T	22	13	18	53
ΣX	364	276	327	967
ΣX^2	7248	4424	5725	50,687*

* This number is the sum of the squared totals in the column above, that is, $(85)^2 + (29)^2 + (22)^2 + \cdots + (53)^2$.

these acuity test objects which the subjects could read under the three experimental conditions. One of the effects of anoxia at high altitudes is a dimming of vision. This test was specifically designed to provide some quantitative evidence of this dimming of vision.

Variations in anoxia were produced by putting the subjects in an altitude chamber from which air could be pumped to simulate atmospheric conditions at various altitudes. Exhausting the air from such a chamber is noisy and obvious. This in itself often has an important psychological effect on the people in it. The subjects know that air is being exhausted from the chamber and they know something about the symptoms that occur at high altitudes. Thus, the mere fact of their sitting in the chamber, with all of the attendant noise and activity, occasionally produces symptoms simply through the power of suggestion. For this reason, the first condition tested was a simulated altitude of 2,000 feet. The subjects were not told at what altitude they would be tested. They were merely seated in the chamber. The door was sealed and the pumps were started. However, by properly balancing the air intake and outlet the chamber was, after several minutes, brought to an equilibrium at a simulated altitude of only 2,000 feet. After sitting there for several minutes, the visual test was administered.

The next step was to bring the altitude chamber to a simulated altitude of 15,000 feet. After sitting there for several minutes, the subjects took the visual test again. Finally, they put on oxygen masks, breathed 100 per cent oxygen and took the test once more. Notice that in this research plan each subject took the test three times, once under each of the three experimental conditions.

The analysis of variance is shown in Table 16. The total sum-of-squares is computed, by analogy with previous computations, as follows:

$$\Sigma x^2{}_T = 7,248 + 4,424 + 5,725 - \frac{(967)^2}{60}$$

$$= 17,397 - 15,584.82 = 1,812.18$$

The variance between experimental conditions is computed in the same way as it was in the previous example:

$$\Sigma x^2{}_{EC} = \frac{(364)^2}{20} + \frac{(276)^2}{20} + \frac{(327)^2}{20} - 15,584.82 = 195.23$$

This time, however, we have an additional source of variation not contained in the previous examples. This is the variation between individuals. It is computed by working with the totals for individuals in the extreme right-hand column. The computation is as follows:

$$\Sigma x^2{}_{BI} = \frac{85^2 + 29^2 + 22^2 + 55^2 + \cdots + 53^2}{3} - 15{,}584.82$$

$$= \frac{50{,}687}{3} - 15{,}584.82 = 1{,}310.85$$

(Where does the 3 come from in these computations?)

The final term, the error term, is obtained by subtraction from the total. There is a way of computing this error term independently in a

TABLE 16. *Analysis of variance for the data in Table 15*

Source of variation	Sum of squares	Degrees of freedom	Estimate of variance	F-ratio
Between individuals	1,310.85	19	68.99	
Between experimental conditions	195.23	2	97.62	$\dfrac{97.62}{8.06} = 12.11$
Residual (experimental error)	306.10	38	8.06	
Total	1,812.18	59		

table of this sort, but the computations are so lengthy and complicated that we shall not discuss them here. If you are interested, consult a textbook of advanced statistics.

The F-ratio evaluates the significance of the differences between experimental conditions. It is entered into the appropriate probability tables with 2 and 38 degrees of freedom. F-tables show that the differences are highly significant, since the probability that they could have occurred by chance is less than 0.1 per cent.

One of the most important things to note about this research design is that it is usually much more sensitive than the kind illustrated in Tables 13 and 14. If you will compare Tables 14 and 16, you will see that in Table 16 we were able to extract a sum-of-squares which we were not able to extract in Table 14. This is the sum-of-squares which can be identified with variations between individuals. Moreover, note

that the sum-of-squares for individuals is taken out of what would normally have been the residual sum-of-squares. For example, if the 60
scores in Table 15 had each been obtained on a different subject, the
residual sum-of-squares in Table 16 would have been 1,616.95 (1,310.85
+ 306.10) with 57 (19 + 38) degrees of freedom. This would have given
an estimated residual variance of 1,616.95/57 or 28.37. The *F*-ratio
would then be 97.62/28.37 = 3.44 and would have been barely significant at the 5 per cent level. The design in Table 15 is more sensitive
than the one in Table 13, because we can identify another source of
variation and so remove it from the unaccounted sources of variation
which go to make up the error term. The design illustrated in Table 15
is an illustration of the type in which each subject serves as his own
control—about which we shall say more in Chapter 5.

On interactions. The error term computed in this example also merits
a little discussion, since we shall have to deal with a similar one in the
next example. This error term is technically an *interaction* between subjects and experimental conditions. An interaction measures discrepancies between patterns of measurements in one variable when these
patterns of measurements are computed for different values of a second
variable. Let's see what this means. Table 15 shows that, on the average, subjects received a high score at 2,000 feet; received a lower score
at 15,000 feet breathing air; and then received a higher score at 15,000
feet breathing oxygen. For example, *Q* received scores of 19, 13, 19.
The first and third scores are equal and the second score is 6 less than
the other two. If *R* had also received scores of 15, 9, 15, the *pattern*
of his scores would agree with those of *Q*. Similarly, if *L* had received
scores of 20, 14, 20, the pattern of *his* scores would agree with *Q*'s. I
keep emphasizing the word *pattern* because, as you see, the total number of items reported correctly by each of these subjects is different.
These differences in over-all performance do not concern us because
they are taken out when we extract the sum-of-squares for the differences due to subjects.

If every subject in this experiment had had the same pattern of
scores—the first and third scores equal and the second 6 less than the
first and third—there would be no interaction between subjects and
experimental conditions. Inspection of the data in Table 15, however,
shows that every subject did not conform to this pattern. For example,
although the first and third scores for *R* are the same, his second score is
only 4 less than the other two. Subject *A*, in fact, earned his highest
score on the second test—a pattern which is atypical. Our error term,

the interaction, then is actually a measure of the discrepancies in the way different subjects behaved with respect to the three experimental conditions.

The final point about interactions concerns the calculation of the number of degrees of freedom that get assigned to them. This is always the product of the number of degrees of freedom associated with each of the basic variables entering into the interaction. In Table 16, for example, the interaction of subjects times experimental conditions has 38 degrees of freedom associated with it because there are 19 degrees of freedom associated with subjects and 2 degrees of freedom associated with experimental conditions. Nineteen times two equals 38. Be sure to check this rule in Table 18 and in all the other analysis of variance problems throughout the rest of this book.

A subjects-by-treatment design with replication. In the experiment on the stoves, I mentioned that each subject had received a total of 80 trials on each stove. The data in Table 13 show only the results obtained on the last trial. They were deliberately selected in this way to simplify the statistical analysis for purposes of exposition. Now let us see what happens when we add the data from some other trials. To keep

TABLE 17. *Reaction times (in seconds) earned by 45 subjects tested on the three stoves in Figure 2 (unpublished data from Chapanis[28])*

	Subjects	Trial 78	Trial 79	Trial 80	Sums
	A	0.64	0.49	0.52	1.65
	B	0.56	0.58	0.59	1.73
	C	0.47	0.39	0.35	1.21
	D	0.80	0.51	0.38	1.69
	E	0.58	0.51	1.35	2.44
	F	0.65	0.53	0.52	1.70
	G	0.66	0.57	0.60	1.83
Stove 1	H	0.88	0.75	0.63	2.26
	I	0.75	0.48	0.68	1.91
	J	0.58	0.55	0.49	1.62
	K	0.67	0.52	0.54	1.73
	L	0.47	0.43	0.41	1.31
	M	0.55	0.43	0.46	1.44
	N	0.49	0.45	0.38	1.32
	O	0.78	0.69	0.57	2.04
	ΣX	9.53	7.88	8.47	25.88
	ΣX^2	6.2727	4.2688	5.5783	46.2948*

* See footnote at end of table.

TABLE 17 (CONT.)

	Subjects	Trial 78	Trial 79	Trial 80	Sums
Stove 2	P	0.88	1.20	0.67	2.75
	Q	2.18	1.45	0.73	4.36
	R	0.60	0.49	0.56	1.65
	S	0.61	0.52	0.64	1.77
	T	0.45	0.44	0.92	1.81
	U	0.83	0.87	0.96	2.66
	V	0.89	0.57	0.60	2.06
	W	0.60	0.40	1.20	2.20
	X	0.76	0.62	0.81	2.19
	Y	0.53	0.49	0.74	1.76
	Z	1.05	0.83	0.93	2.81
	A'	0.46	0.40	0.69	1.55
	B'	0.70	0.63	0.72	2.05
	C'	0.71	0.44	0.86	2.01
	D'	0.46	0.50	0.73	1.69
	ΣX	11.71	9.85	11.76	33.32
	ΣX^2	11.6807	7.8023	9.6086	81.1538*

	Subjects	Trial 78	Trial 79	Trial 80	Sums
Stove 3	E'	0.47	0.52	0.41	1.40
	F'	0.44	0.50	0.43	1.37
	G'	1.05	0.81	0.84	2.70
	H'	0.73	0.82	0.66	2.21
	I'	0.39	0.51	0.43	1.33
	J'	0.62	0.65	0.68	1.95
	K'	0.52	0.60	0.50	1.62
	L'	0.59	0.51	0.51	1.61
	M'	0.74	0.83	0.68	2.25
	N'	0.76	0.54	0.62	1.92
	O'	0.64	0.80	1.05	2.49
	P'	0.59	0.55	0.52	1.66
	Q'	0.60	0.56	0.98	2.14
	R'	0.68	0.74	0.72	2.14
	S'	0.61	0.60	0.82	2.03
	ΣX	9.43	9.54	9.85	28.82
	ΣX^2	6.2823	6.2958	7.0225	57.7836*
	$\Sigma\Sigma X$	30.67	27.27	30.08	88.02
	$\Sigma\Sigma X^2$	24.2357	18.3669	22.2094	185.2322*

* Each of these numbers is the sum of the appropriate squared totals in the column above. For example, $81.1538 = 2.75^2 + 4.36^2 + 1.65^2 + \cdots + 1.69^2$.

the arithmetic down to simple proportions, let us consider only the data for the last three trials. These are shown in Table 17. Note that the entries in Table 13 are exactly the same as those for Trial 80 in Table 17. The effect of adding the two additional trials is to provide us with a replication—as the statisticians call it—a double repetition of the experiment.

TABLE 18. *Analysis of variance for the data in Table 17*

Source of variation	Sum of squares	Degrees of freedom	Estimate of variance	F-ratio
A. Between individuals	4.3551	44		
1. Between stoves	0.6241	2	0.3120	$\dfrac{0.3120}{0.08883} = 3.51^*$
2. Between individuals within stoves	3.7310	42	0.08883	
B. Between trials	0.1467	2	0.07335	$\dfrac{0.07335}{0.03320} = 2.21$
C. Interaction: Trials × individuals	2.9212	88	0.03320	
1. Interaction: Trials × stoves	0.1108	4	0.02770	$\dfrac{0.02770}{0.03346} = 0.83$
2. Interaction: Trials × individuals within stoves	2.8104	84	0.03346	
D. Total	7.4230	134		

* Significant at the 5 per cent level.

The simplest way to attack the analysis of such a set of data is to start by ignoring the fact that these subjects were tested on different stoves. Think of this as a relatively simple table with 3 trials across the top and 45 subjects running down along the side. If you do this then Table 17 is exactly analogous to the research design illustrated in Tables 15 and 16. The three trials in Table 17 correspond, by analogy, with the three experimental conditions in Table 15. The 45 subjects in Table 17 correspond, by analogy, with the 20 subjects in Table 15.

Again, by analogy with the data in Table 16, the four principal components of variance are shown by the capital letters A, B, C, and D in Table 18. The total sum-of-squares is computed in this way:

$$24.2357 + 18.3669 + 22.2094 - \frac{(88.02)^2}{135}$$

$$= 64.8120 - 57.3890 = 7.4230$$

The sum-of-squares between individuals (all 45 individuals) is:

$$\frac{(1.65)^2 + (1.73)^2 + (1.21)^2 + \cdots + (2.03)^2}{3} - 57.3890$$

$$= \frac{185.2322}{3} - 57.3890 = 61.7441 - 57.3890 = 4.3551$$

The sum-of-squares between trials is:

$$\frac{(30.67)^2 + (27.27)^2 + (30.08)^2}{45} - 57.3890 = 0.1467$$

The interaction (or error) term, C, is obtained by subtraction as in Table 16.

Now, however, we must take account of the fact that the 45 subjects were in fact tested on different stoves. In short, part of the variation between subjects (in A of Table 18) arises from the variation between stoves and part from the variation between individuals tested on the same stove. The sum-of-squares due to the differences between stoves is:

$$\frac{(25.88)^2 + (33.32)^2 + (28.82)^2}{45} - 57.3890 = 0.6241$$

The sum-of-squares due to variations between individuals who were tested on the same stove is computed as follows:

For stove 1: $\dfrac{(1.65)^2 + (1.73)^2 + (1.21)^2 + \cdots + (2.04)^2}{3} - \dfrac{(25.88)^2}{45}$

$$= \frac{46.2948}{3} - 14.8839 = 15.4316 - 14.8839 = 0.5477$$

For stove 2: $\dfrac{81.1538}{3} - \dfrac{(33.32)^2}{45} = 2.3797$

For stove 3: $\dfrac{57.7836}{3} - \dfrac{(28.82)^2}{45} = 0.8036$

For all stoves: $0.5477 + 2.3797 + 0.8036 = 3.7310$

Note that the sum-of-squares for the variation "between stoves" and the sum-of-squares for the variation "between individuals within stoves" add up to the sum-of-squares for the variation "between individuals."

The sum-of-squares arising from the variation between trials cannot be subdivided any further, but the interaction term in C of Table 18 can be subdivided. The simplest way of showing the computation of one component is to arrange a sub-table as in Table 19. In this table, we ignore the fact that there are different individuals in the experiment and consider only the variation between stoves and trials. The entries

TABLE 19. *This table was prepared from Table 17 to help in computing the stoves $\times$ trials interaction*

	Trial 78	Trial 79	Trial 80	Sums
Stove 1	9.53	7.88	8.47	25.88
Stove 2	11.71	9.85	11.76	33.32
Stove 3	9.43	9.54	9.85	28.82
Sums	30.67	27.27	30.08	88.02

in Table 19 are made up of the appropriate subtotals in Table 17. The total sum-of-squares for the data in Table 19 is:

$$\frac{(9.53)^2 + (7.88)^2 + (8.47)^2 + (11.71)^2 + \cdots + (9.85)^2}{15}$$

$$- 57.3890 = 0.8816$$

From this total we subtract the sum-of-squares for stoves and the sum-of-squares for trials (both of which are already computed) thus:

$$0.8816 - 0.6241 - 0.1467 = 0.1108$$

The remainder is the interaction of stoves times trials. It tells us whether the pattern of performances on the three trials was the same for all three stoves.

The last component of the analysis in Table 18 is obtained by considering the data for each stove separately. For example, if we take only the data for stove 1, we have a set of data arranged exactly like those in Table 15. We can first compute a total sum-of-squares:

$$6.2727 + 4.2688 + 5.5783 - \frac{(25.88)^2}{45}$$

$$= 16.1198 - 14.8839 = 1.2359$$

a sum-of-squares between trials:

$$\frac{(9.53)^2 + (7.88)^2 + (8.47)^2}{15} - 14.8839 = 0.0932$$

and a sum-of-squares between individuals:

$$\frac{(1.65)^2 + (1.73)^2 + (1.21)^2 + \cdots + (2.04)^2}{3} - 14.8839$$

$$= \frac{46.2948}{3} - 14.8839 = 15.4316 - 14.8839 = 0.5477$$

The remainder $1.2359 - 0.0932 - 0.5477 = 0.5950$ is the interaction of subjects times trials on stove 1 only.

The two comparable interaction terms for stoves 2 and 3 are 1.8823 and 0.3331. Together they add up to the interaction of trials-times-individuals within stoves, as shown in Table 18. Note again that the sum-of-squares for the interaction of trials-times-stoves and the sum-of-squares for the interaction of trials-times-individuals within stoves add up to the sum-of-squares for the interaction of trials-times-individuals.

To test the differences between stoves we must test the variance between stoves, 0.3120, by its own error term, 0.08883. This F-ratio is equal to 3.51 and is significant at the 5 per cent level. This means that the reaction times obtained on the different stoves are significantly different. To test the variance between trials, we divide 0.07335 by 0.03320. The resulting F of 2.21 is not significant. This means that, as far as we can tell from these data, the variation between trials is no larger than would be expected by chance. Finally, to test the interaction of trials-times-stoves, we must divide 0.02770 by 0.03346. The resulting F of 0.83 is not significant, indicating that the three stoves gave the same pattern of results on the three trials.

Other statistical designs. This is as far as we shall go into the mechanics of analyzing data for the significance of differences. These examples illustrate some of the simpler and more useful types of research designs used in human engineering work. If you master the basic principles of these analyses you should be able to handle easily the examples discussed later in Chapters 5 and 8. All of these together will give you a substantial battery of useful research designs for practical work in this area. For types of statistical analyses not discussed here you should consult one of the more advanced textbooks on this subject.

SOME GENERAL REMARKS ON EVALUATING
THE STATISTICAL SIGNIFICANCE
OF DIFFERENCES

To conclude this discussion of why we should be concerned with statistics, let us look at a concrete illustration. Here is an industrial research report which describes two ways of doing a task. Let's call them *A* and *B*. Method *A* gave an average value of 8.97 seconds; *B* of 9.30 seconds. *A* is shorter than *B*, and this is all the author said about it. But what are we to conclude: Is *A* *really* shorter than *B*? What is our confidence in the relative order of the two? Since this is an old study and the raw data are no longer available, it is impossible to say. We know that people vary—they vary from person to person and from day to day. If the data of this experiment were distributed like those in the upper half of Figure 25, we would probably agree that the difference between the two methods is significant, that is, that the difference is not the result of chance factors. If, on the other hand, the data were like those in the lower half of Figure 25, we would be much more likely to conclude that this is a chance difference.

As you have seen, statistics give us a precise way of quantifying such judgments. The mean values in the upper and lower halves of Figure 25 are exactly the same. However, if you will apply formula [20] to the two sets of data, you will discover that a difference as large as that shown in the upper half of the data could have occurred by chance less than once in a thousand times. This is so unlikely an event that we conclude that the difference is genuine—not the result of chance factors. The same sort of analysis applied to the data in the lower half will show that the same difference could have occurred by chance more often than five times in a hundred. These odds are not very good, and most experimenters would conclude that the difference is not dependable.

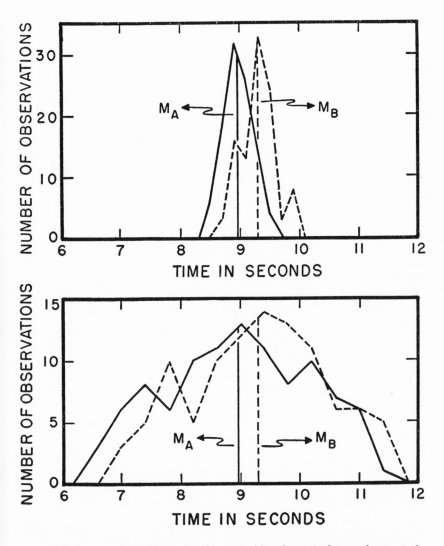

FIG. 25. Average times (shown by the vertical lines) required to perform a task with two different kinds of equipment, *A* and *B*. The statistical significance of such a difference can be evaluated only in terms of the variability in the data. If the variability is small, as in the chart above, we may have confidence in the significance of the difference between these averages. If the variability is great, as in the chart below, we conclude that the difference between the averages for *A* and *B* may be the result of chance. Statistics enable us to attach probabilities to such statements.

This is only one example, but it is not hard to find many more like it in any number of studies done in industry. What's the moral of the story? It is simply this: The experimenter's obligations do not end when he has finished running an experiment. He also has to communicate his findings, and to communicate them in a way which will be useful to other people. In general, this means that the experimenter is obligated to study his data thoroughly, to subject them to a complete statistical summary, and to evaluate them for their statistical and practical significance.

On a cost-accounting basis, the most expensive part of any experiment is generally in the design and construction of the apparatus. Testing subjects is frequently the second most expensive part of the experiment (although in certain long, drawn-out, learning experiments, this may be the most time-consuming, and expensive, part of the experiment). Analyzing the data of an experiment is typically the least expensive part of the whole procedure. Why spend months building apparatus, weeks or months testing subjects, and then skimp with a few hours of superficial statistical analysis and interpretation of the results? Many old experimental data, costing literally tens of thousands of dollars, are now almost useless because they were not analyzed and reported in sufficient detail for us to evaluate them properly. As long as you work with equipment where stability is high and variability low, simple averages *may* be sufficient. But when you work with people where variability is high, statistical methods are indispensable for the proper evaluation of the data.

EVALUATING THE PRACTICAL SIGNIFICANCE OF EFFECTS

Many experimenters stop when they have answered the question of the statistical significance of an effect; they seldom consider the other part of the problem—its practical significance. Yet in most human engineering work the second question is more important than the first. There are really two parts to the question of practical significance, one statistical, the other a matter of engineering and cost accounting. In this section we shall look only at the statistical part of the problem.

From the statistical standpoint what we want to know when asking about practical significance is this: "Does the effect amount to anything?" Psychologists typically consider that an experiment has reached a satisfactory conclusion if they can demonstrate an effect which could not have occurred by chance. Actually, statistical tests enable you to do more than this. Once you have established that there is a statistically significant difference between the averages obtained, let's say, for equipment *A* and *B*, you can then go on to make an estimate of how large the *true* difference is between these two equipments. This estimate of the true difference is expressed in probability terms, and it gives us boundary conditions (technically called fiducial or confidence limits): It tells us that the true difference between *A* and *B* is almost certainly as large as *X*, and almost certainly not larger than *Y*. You can readily see why this kind of statement is so important for human engineering work. By having such estimates to look at, the engineer can then decide whether the gain to be anticipated by changing all equipment *A*'s to *B*'s will offset the cost of doing it. Without such estimates, the engineer lacks the necessary information to enter into his cost equations.

Confidence limits for the data in Table 11. To illustrate this procedure let us compute estimates of the true difference between the two groups in Table 11. Tables of the *t*-distribution show us that a *t* of 2.02 is significant at the 5 per cent level. This value of 2.02 is multiplied by the standard error of the difference, 2.41, computed on page 124. The product, 4.9, is then added to and subtracted from the mean difference of 9.4. We see that 9.4 minus 4.9 equals 4.5; 9.4 plus 4.9 equals 14.3. This tells us that the probability is 19 in 20 that the true difference between the two groups is between 4.5 and 14.3. Since a *t* of 2.70 is significant at the one per cent level, similar computations tell us that the chances are 99 in 100 that the true difference is between 2.9 and 15.9.

Having made an estimate of the probable size of the true difference, one can then go on to ask an additional statistical question: "How much of the total variability does this effect account for?" There are many statistically significant effects which, for all practical purposes, are worthless. They are worthless because they account for so little of the total variability in the phenomena we are studying. For example, one study has found a correlation between height and intelligence amounting to about $+0.10$. With a big enough sample of people this is a statistically significant value. In general, taller people tend to be more intelligent than shorter ones. But from a practical standpoint this is virtually useless. A correlation of 0.10 accounts for only one per cent of the variability, that is, one per cent of the variability in intelligence can be accounted for by variations in height. This is such a small value it is not worth bothering with. There are much larger and more important predictors of intelligence. For a discussion of the evaluation of the practical significance of errors in man-machine systems see the article by Chapanis.[25]

STATISTICS AND EXPERIMENTAL DESIGN

By now it should be apparent that statistics and experimental design go hand in hand in a very real sense. Although only a few basic principles of statistical analysis were used in the examples in Tables 11 through 17, the precise form of an analysis depends on the way the data were collected in the first place. For each of the sets of data in Tables 11, 13, 15, and 17 there was an appropriate form of statistical analysis. However, it does not follow that every set of data can be analyzed. Herein lies the trouble. Statisticians are continually being approached by experimentalists with the request, "Here are some data I collected on a problem. Will you help me to calculate some statistics on them?" Sometimes it turns out that no meaningful statistical analysis is possible because the experiment was not designed properly in the first place.

Before the experimenter starts collecting any data, at the time he is still planning how to run his tests, he should also be planning the exact form of the statistical analysis he will do on the data when he gets them. This kind of planning will protect the investigator in two ways: (a) he will be sure, in advance, that he will be able to summarize and evaluate his data; and (b) he will be able to spot weaknesses, or fallacies, in his experimental design before he starts running tests.

In the latter connection, one of the most important things to insure in planning an experiment is that the design include within it some procedure for measuring the effects of random or residual variability. We cannot compute the statistical significance of effects unless we know the size of the random error. In human engineering work the best way to get estimates of random error is to repeat measurements by testing different subjects under each of the conditions (as was done in Tables 11 and 13) or by testing the same subjects in each experimental condition (as was done in Tables 15 and 17). The next chapter will consider this matter further.

SOURCE MATERIAL ON STATISTICS

Despite their importance for human experimentation, we shall not say any more than this about the statistical methods. There are a large number of good textbooks available on the subject (see items 1, 36, 41, 44, 45, 56, 68, 73, 81, 89, and 101 in the bibliography), and the reader who wants to become more familiar with these techniques will find it instructive to consult them before beginning any large-scale program of human engineering studies.

| ***The experimental method***

As in the other sciences, the cream of research techniques is the experimental method. Basically, the kinds of experiments you do with people are similar to the kinds you do in the other sciences. The details differ, the controls are more involved and sometimes impossible to apply, but the logic and point of view underlying all experimental methods are essentially the same.

ON THE NATURE OF AN EXPERIMENT

Despite the importance and universality of the experimental method in science, it is difficult to find two scientists who will agree on how to define it. As a point of departure, however, we can say that in human research an experiment seems to have these important features: It is a series of *controlled observations* undertaken in an *artificial situation* with the *deliberate manipulation of some variables* in order *to answer one or more specific hypotheses*. The four separate characteristics italicized in this definition provide the clues as to exactly why experiments are so useful.

Controlled observation. Chapters 2 and 3 described a number of more-or-less sophisticated observational techniques, but even the most elaborate series of observations, by itself, does not make an experiment. There are some other critical ingredients that make the difference. When the scientist sets up an experiment he plans, controls and describes all of the circumstances surrounding his tests. Not only does this give him greater control over the course of events but it also enables him to set up conditions so that he can repeat the experiment if he wants to. Direct observations of man-machine systems practically never allow the investigator to repeat observations under identical conditions.

More important than this, however, is the fact that when an experiment is done properly and described adequately, anyone else can repeat it. Experiments are publicly verifiable. This repeatability of experiments—in your laboratory or mine, in the United States, South America, or Russia—is what gives scientific laws their dependability.

The artificial situation. Toward the end of this chapter we shall see that there are various degrees of realism possible in an experiment. By and large, however, experiments are artificial situations contrived by the investigator. Experiments are artificial in two senses. (a) They are situations which do not crop up naturally but are deliberately created by the experimenter's art. (b) People often behave artificially in experiments, that is, the subjects in an experiment rarely become so completely engrossed in the experimental task that they behave naturally.

One of the most important reasons for creating such artificial situations is that the experimenter can make an event happen at a certain time and place. This means, among other things, that he is prepared to make accurate observations of the event because he knows when to expect it. Another reason why experiments are done in an artificial setting is to allow the experimenter to control and manipulate the variables which might affect the outcome of his observations.

Deliberate manipulation of variables. Another important difference between an experiment and mere observation is that in an experiment the research man can systematically vary conditions and note the concomitant variations in results. He does not have to take things as they come, as in the critical-incident technique or the study of near-accidents, for example. This also means that the investigator can, if he wishes, try combinations which do not occur, or have not yet occurred in real life. Scientists are able to do experiments on many human problems of space flight, although no one has yet (in 1958, that is) gotten there.

The deliberate manipulation of variables also enables us to get at "causes" with greater certainty than we can possibly get from mere observation. One of the troubles with observations on man-machine

systems is that they are so complicated. If we look at the events leading up to the one we are interested in, we find many things which
might be the cause of what we observe. How can we untangle them?
It is impossible to do this merely by observing. But an experiment is
an elegant attack on this problem because it enables us to eliminate
factors, one by one, to vary their strengths, or to combine them in predetermined ways. If we think that any one of factors A, B, C, D, or E
might be responsible for a certain kind of behavior, we can often nail
down the important one by deliberately eliminating these one at a
time. Sometimes much more involved procedures are necessary to get
at causes, but the essential point is that an experiment is a powerful
ally in our struggle against nature to find out what causes what.

Testing specific hypotheses. The final point about experiments is that
they are usually undertaken with a fairly definite problem in mind.
Observations of the type described in Chapters 2 and 3 are usually
much less specific—the investigator is not quite sure what he is looking for, or even what he hopes to find out. He is merely probing. In an
experiment, however, he has usually focused his attention on some
specific questions or hypotheses. Having such a point of focus means
that he can sharpen his whole procedure and plan efficiently to discover the answers to the questions he has posed. This is at the heart
of the problem of experimental design, a problem which will be discussed at greater length in the next section.

On the role of observation in experimentation. In this discussion of
experiments you may have formed the impression that the observational methods we talked about in Chapters 2 and 3 are of practically
no use. This is far from the case. Observation is the primary source of
the ideas we need for our experiments. Observations on highway
accidents suggest that certain signs are not very visible. This has provided the stimulus for many carefully controlled experiments on the
legibility of various kinds of numerals, letters, and symbols, and on
other factors that contribute to the legibility of signs in general.
Observations on difficulties encountered in using certain types of
controls have led to many useful experiments on the optimum sizes of
controls, the gearing ratios between controls and displays, and the
relationships between directions-of-movement in controls and displays. In general, to do good experiments you must have some ideas
about the variables that need to be explored. These you often get from
casual or systematic observations of the kinds discussed earlier.

THE DESIGN OF EXPERIMENTS

The formal pattern or plan for collecting observations is called the design of an experiment. An experimental design should always be constructed before the investigator actually starts collecting data. The principal ingredients of such a plan are (a) the identification of the variables to be investigated, (b) some decisions about the kind and number of experimental subjects to be used, and (c) a carefully arranged scheme governing the order in which experimental trials are to be run. It is this last point with which we shall be primarily concerned in this section.

SOME GENERAL PRINCIPLES OF EXPERIMENTAL DESIGN

Before getting into the technical details of setting up experiments to test various hypotheses, we need to look at some important general principles underlying experimentation in human engineering.

The design should yield a measure of the random error in the experiment. In the more exact sciences experimental errors are often so small that one need not be greatly concerned about them. But in human research one must always be sure to build into an experimental design some procedure for getting an estimate of the experimental error. This estimate of experimental error must come from the *same experiment as is used to test the primary factors you are interested in.*

Suppose, for example, that we are concerned with testing two kinds of handles for escape hatches—call them X and Y. In particular we want to find out whether it is possible for a man to exert more force on one or the other of these handles. We take an experimental subject, station him in front of X, and measure the amount of force he can exert on it. We then repeat the observation with handle Y. Suppose the data come out this way:

Handle $X:$ 4,000 inch pounds
Handle $Y:$ 3,750 inch pounds

What are we to conclude? Is X better than Y? We cannot say, and there is no way to find out from these data. When we work with human subjects, one thing we can be sure about is that they will vary from one occasion to the next. Even if we were to test the same man on the same handle several times in succession, the chances of getting *exactly* the same reading twice in a row are very remote. This inherent variability in our test material is a major factor contributing to the experimental error in all human experiments. The experimental, residual, or random error refers to those variations in outcome which we cannot control and cannot account for. Until we have some estimate of the size of this error we cannot evaluate an experimental outcome.

Generally one provides for estimates of random error in an experiment by what statisticians call *replication*, or, to use a less exact but more familiar word, by repetition. There are in general two ways of doing this: (a) we may use more than one subject; or (b) we may use more than one trial. The first of these techniques was used in the experiment as shown in Table 11 of the preceding chapter. The second can be illustrated by continuing with our example on handles.

Suppose that we were to have our experimental subject make two trials with each of the controls X and Y. If the data came out this way:

	Trial 1	Trial 2	Mean
Handle X:	4,000 inch pounds	3,700 inch pounds	3,850 inch pounds
Handle Y:	3,750 inch pounds	3,950 inch pounds	3,850 inch pounds

we are in a much better position to evaluate our initial experimental outcome. We can now see that the experimental error, the average difference between readings obtained under identical experimental conditions ($\frac{1}{2}\{[4,000 - 3,700] + [3,950 - 3,750]\}$), is 250, or exactly as large as the difference we found on Trial 1. Under these circumstances, you can see that you should not have placed any confidence in the difference between X and Y. In fact, the averages for the two trials together are exactly the same.

On the other hand, if the data had come out this way:

	Trial 1	Trial 2	Mean
Handle X:	4,000 inch pounds	3,950 inch pounds	3,975 inch pounds
Handle Y:	3,750 inch pounds	3,715 inch pounds	3,732.5 inch pounds

you would undoubtedly be much more inclined to trust the difference between X and Y. The mean difference between the outcomes obtained with these two controls is much greater than the experimental error. Exact tests of the confidence we should place in such an outcome have already been discussed in the preceding chapter. Applying such a test to the data above will show that the mean difference between handles is statistically significant. A little reflection, however, will show that this is a trivial finding because it tells us that there is a difference between the two handles, but only *for the one subject we tested*. As a general rule then we should try to get estimates of random error by replicating subjects and not by replicating trials on the same subject.

Trials should be counterbalanced to avoid systematic bias. The preceding section was primarily concerned with the more-or-less random, or erratic, changes which characterize human performance in an experiment. There are also, however, some fairly regular changes we must guard against to avoid what statisticians call *systematic bias* in the data. The nature of this problem can be demonstrated by continuing with the example used above. How should we schedule the trials on the two handles? Suppose that we had run the tests in the following order and had obtained results as before:

Test 1:	Handle X, Trial 1	4,000 inch pounds
Test 2:	Handle X, Trial 2	3,950 inch pounds
Test 3:	Handle Y, Trial 1	3,750 inch pounds
Test 4:	Handle Y, Trial 2	3,715 inch pounds

What do we conclude? One thing which is immediately apparent is that the readings decline fairly consistently throughout the tests. Since these tests involve muscular force, we can at least suspect that the subject was getting tired. This progressive decrease in readings throughout the four tests also makes it much more difficult to interpret the average difference obtained with the two handles. There may very well be a genuine difference between X and Y, but you will have to admit that this interpretation is clouded by the fact that the tests with Y were all run *after* those with X.

It would be nice if people were relatively stable materials upon which to do experiments. Unfortunately this is not the case. Human beings do not stand still, and the person you end up with is by no means the same person you began your experiment with. He has changed because, among other things, he has learned something in the meantime. He may also have become fatigued, bored, or even more interested in your experiment as it progressed. Any or all of these mean that you must always expect, and plan for, regular changes in human behavior during the course of an experiment. Unfortunately, you cannot always predict the ways in which behavior will change during an experiment. If the subjects learn the experimental routine, or learn how to operate the experimental apparatus more efficiently, they may show continual improvement in performance. On the other hand, such learning may be completely masked by progressive fatigue throughout the experiment.

One way to guard against bias in experimental outcomes from such changes is to counterbalance trials. In the example above, suppose that we wanted to make four tests each with X and Y. A counterbalanced series of trials would be the following: $XYYX\ YXXY$. With this type of schedule, any progressive changes in behavior will be averaged out over the series. Other examples of counterbalancing in more complex experimental designs will be illustrated later in this chapter.

To show what sometimes does happen, consider the following set of data. An investigator wanted to find out whether tilting a keyset would improve the accuracy with which operators could key out numbers. Two girls keyed out numbers one hour a day for eight days. During these trials, the keyset lay flat on the operating board. On the ninth day, the keyset was tilted to 20°, and the operators used the keyset in this position for the next three days. The results reported were that, on the average, operators had a keying time of 9.0 seconds for the 0° tilt, and 7.7 seconds for the 20° tilt. Off hand, it looks as though tilting the keyset was an improvement. However, this investigator was astute enough to recognize that this could be the result of progressive learning. When the data are plotted according to day of test, the results are as appear in Figure 26. You will have to agree that, when they are expressed this way, there is slender evidence for anything other than continued learning throughout the entire series of trials.

Although, in this case, the data were correctly interpreted before any false conclusions were drawn, the essential point is that an experiment like this should never have been designed in the first place. In working with human subjects, one must always go on the assumption that there will be changes in behavior during the course of the experiment. There are good and workable techniques for taking care of such problems, and they should be built automatically into human experiments.

When you do not counterbalance, randomize. As an alternative to counterbalancing, the best general rule to follow is that the assignment

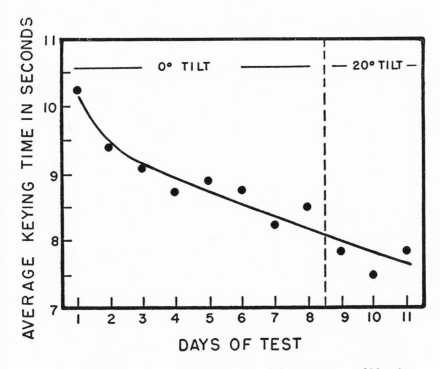

FIG. 26. The results of a study on keysets showed that operators could key faster when the keyset was inclined at 20°. The curve shown here was obtained by plotting average keying time for two subjects against successive days of test. Looking at it in this way it is impossible to separate the effects of tilting the keyset from the effects of learning. Proper design of the experiment would have obviated this difficulty.

of trials and subjects should be made by using a table of random numbers. This is the best way of guarding against any of a large number of biases which can creep into human experiments. The use of randomization will be demonstrated and amplified in the material which follows.

There are times, of course, when the experimenter is interested in the progressive changes of behavior which occur in an experiment. Learning experiments, or experiments on fatigue are of this type and are the proper object of scientific inquiry. Obviously one would not counterbalance or randomize trials in such instances.

Do not confound variables. In statistical terms two variables are said to be confounded when they cannot be untangled or isolated. Let's see what this means in terms of our example with the two handles. Suppose that we had tested two different subjects on the two handles and followed this experimental schedule with the results shown:

	Trial 1	Trial 2	Mean
Subject *A*, Handle *X:*	4,000 inch pounds	3,950 inch pounds	3,975 inch pounds
Subject *B*, Handle *Y:*	3,750 inch pounds	3,715 inch pounds	3,732.5 inch pounds

What can we conclude from these data? There is clearly a difference between the results obtained on the two handles and the average difference is much larger than the experimental error. But to what do we attribute the difference? To the controls? Or to the subjects? We cannot untangle these two possibilities, because *A* was the only subject tested on Handle *X*, and *B* was the only subject tested on Handle *Y*. Clearly a much better experimental design is one in this form:

	Handle *X*	Handle *Y*
Subject *A*	Trial 1 Trial 4	Trial 2 Trial 3
Subject *B*	Trial 2 Trial 3	Trial 1 Trial 4

This design untangles subjects from handles, it provides us with an estimate of experimental error, and it counter-balances for progressive changes which might occur during the experiment.

It is easy to see the difficulties introduced by confounding in a simple example like the above. Yet it is surprising to see how often confounding is unwittingly introduced into real-life experiments. I recall one experiment done a good many years ago to test the insulation properties of two kinds of arctic sleeping bags—call them *F* and *G*.

The tests were made in a cold room by having a squad of men sleep in type *F* sleeping bags on the first night. Thermocouples attached to the men allowed the investigator to record skin temperatures throughout the night. On the second night another squad of men slept in type *G* sleeping bags under identical temperature conditions. The results showed that on the average the men who used the type *G* sleeping bags were warmer than those who had used the type *F* bags. However, the statistician who was to analyze the data found, in talking to some of the subjects, that all the men had slept nude the first night, and had worn heavy woolen underwear on the second!

Another experiment run many years ago had been designed to test the shock of parachute openings on the human body. These were realistic tests involving parachute drops from a four-motored aircraft at various altitudes and at various speeds. The entire series of tests must easily have cost several hundred thousand dollars. One of the findings was that parachutes dropped at an altitude of 25,000 feet produced a significantly greater opening shock than those dropped at 35,000 feet. Upon closer inspection, however, I discovered that *all* of the parachutes dropped at 25,000 feet were nylon parachutes with a 30-foot canopy, whereas all those dropped at 35,000 feet were silk parachutes with a 26-foot canopy!

As an exercise try redesigning the sleeping bag and parachute experiments to get at the relevant variables without contamination.

As with most things in life there are exceptions to the general principle discussed in this section. The sophisticated statistician or experimenter can sometimes make good use of deliberate confounding. By and large, however, the beginner in this business will be well advised to avoid it.

EXPERIMENTS FOR MAKING SIMPLE COMPARISONS

Many human engineering experiments are concerned with rather simple questions of the type: Is this device better than that? Or, is this way of doing something better than that? There are four different ways of designing experiments for such purposes, and each has some advantages and disadvantages. As we go through them, notice how closely they are tied up with statistical procedures for evaluating the experimental outcomes.

Independent groups. The simplest, and crudest, way of setting up comparative experiments is to use two independent groups of subjects —one group being tested under one condition, the other being tested under the other condition. The experiment on the Link Trainer (illustrated in Table 11) is of exactly this type. An important rule to observe in assigning subjects to the two groups is to be sure that the assignment is done in such a way as to avoid bias. Bias can be introduced into an experiment in many strange and unexpected ways, and the simplest way to guard against it is to use a random assignment procedure. If you have a group of 40 subjects available for the experiment, give them the numbers 1, 2, 3, 4, $\cdots$, 40. Then go to a table of random numbers,* pick a starting point at random, and make assignments on the basis of the order in which these numbers appear in the table. For example, I start near the top of the fourth and fifth columns on Page 5 of such a table and read down pairs of columns, looking for numbers between 01 and 40. The first such number I see is 21. Subject 21 is assigned to Group *A*. The next number I come to is 07. Subject 7 is assigned to Group *B*. The next number is 20. He goes to Group *A*. Number 15 goes to Group *B*, number 6 to Group *A*, and so on. When I run across numbers which have already been assigned I ignore them, of course.

It would be difficult to overstress the importance of avoiding any regular or systematic way of assigning subjects to experimental groups. It is even important to avoid making up your own random numbers, because we know that it is difficult for people to make up sets of random numbers even when they try[26]. People who have been in the business of human research for any appreciable length of time can easily give you examples of important biases which have come into experiments because this rule of random assignment was violated. One experimenter, for example, used a large number of volunteer subjects for his experiment and assigned the first *N* subjects to one experimental group, the second *N* subjects to another experimental group. When the experiment was over he found a systematic difference between the two groups which he had not anticipated. The subjects who volunteered first were different from those who were slower to volunteer! Another experimenter assigned subjects whose last names began with the letters *A* through *M* to one experimental group; those whose last names began with *N* through *Z* to another experimental group. Unfortunately, this biased his data because it made a rather clean

---

* Many textbooks of statistics contain abbreviated tables of random numbers (for example Dixon and Massey[41] and Edwards.[44, 45]) More complete tables may be found in Fisher and Yates[48] and Rand Corporation.[93]

division between two basically different nationality groups which were represented among these subjects.

These and still other case studies like them lead to this important principle: Except when you are deliberately assigning trials or subjects according to some systematic procedure, always make assignments by chance, that is, according to tables of random numbers.

If an experiment of the type just described is likely to extend over an appreciable length of time, it is also a good idea to test the two conditions in a counterbalanced or random sequence. The reason for this is that the experimenter himself may bias the data in subtle ways. Running an experiment, like anything else, is a skill which improves with practice. For this reason the first few trials of an experiment will almost certainly not be conducted the same as the last. The experimenter becomes more skilled at reciting the instructions, he has learned to anticipate questions which the subjects might have, and in many other ways he is able to administer the tests more smoothly. This usually makes for a more comfortable working relationship between subject and experimenter which *can* influence the experimental results. Thus, if all the trials under condition *A* were run first, there could conceivably be a difference between *A* and *B* simply due to such inevitable changes in the experimenter himself.

The statistical analysis of an independent-group type of experiment has already been discussed in Chapter 4 (pages 122 ff.).

Matched groups. A somewhat more refined way of designing a simple comparative experiment is to use two groups of subjects who are matched in some relevant characteristic. In an industrial experiment, for example, you might use two groups of workers with the same average level of productivity. In testing the comparative merits of two different radars, to use another example, you get more precision if you can use two groups of operators with the same average ability, or with the same average amount of experience in using radar, or with the same average grades from radar school. You get still more precision if, in addition to matching for *average* ability, you can also match the groups in terms of their variabilities, i.e., the standard deviations of the groups should also be roughly comparable.

In order to use matched groups, you must, of course, have a criterion on which to do the matching. Some criteria which can be used for this purpose are:

a. experience on the job
b. grades earned in training school
c. some measure of performance or skill on the job

Note, incidentally, that this technique has one serious disadvantage—it usually requires a large initial pool of subjects, because it is often necessary to discard some of them in order to match the two groups.

Groups with matched subjects. Still better results are usually obtained from experiments in which individual subjects are matched in the two groups. If one experimental group contains subjects A, B, C, D, etc., the other experimental group would contain subjects A', B', C', D', etc. In this case, however, subject A matches A' in some relevant characteristic, B matches B', C matches C', and so on. This type of experimental design resembles the matched-groups design except that, in this instance, we are matching individual subjects rather than *groups* of subjects.

Each subject serves as his own control. In general, greatest precision results from those experiments in which every subject serves as his own control, that is, in which each subject is tested under both experimental conditions. (The experiment illustrated in Table 15 was of this type.) If we were doing an experiment on the legibility of two different dials, we would test a number of subjects in such a way that each subject was tested on both dials A and B. Note, however, that it would be a good idea to counterbalance by testing half of the subjects in an AB order; the other half in a BA order.

Although this is an extremely efficient type of design, it cannot be used for certain types of experiments—principally those in which experience on condition A (or B) in some way affects the subject so that he cannot, or should not, be tested on the other condition. In the Link Trainer experiment illustrated in Table 11, for example, we could not use the same subjects in both groups. A person can only learn to fly once.

Statistics to match the experimental design. A point to keep in mind is that the statistical treatment of the data depends on the kind of experimental design you use. Equation [20] in Chapter 4 showed the appropriate form of analysis for experiments using two independent groups of subjects.

In the matched-groups design the appropriate formula for testing the significance of the mean difference between two outcomes is the following:

$$t = \frac{M_A - M_B}{\sqrt{\dfrac{\sigma^2_A}{N_A}(1 - r^2_{CA}) + \dfrac{\sigma^2_B}{N_B}(1 - r^2_{CB})}} \qquad [23]$$

where r_{CA} is the correlation between the matching criterion and the scores earned by the subjects under condition A; and r_{CB} is the correlation between the matching criterion and the scores earned by the subjects in condition B. The t is evaluated with $N_A + N_B - 4$ degrees of freedom.

To put this discussion into concrete terms, suppose that we want to compare the legibility of aircraft dials under dim red illumination (condition A) and under ultraviolet light (condition B). To match our two groups of subjects we make up a special pretest of 20 dials and administer this test to all the subjects under white light. On the basis of these pretest results we now select two groups of subjects so that the average reading time for group 1 equals that for group 2. One of these groups is then assigned to condition A; the other to condition B. The correlation between the pretest scores and the scores earned on the experimental trials by the subjects in group A is r_{CA}; while r_{CB} is the correlation between the pretest scores and the scores earned on the experimental trials by the subjects who were tested under condition B.

In the matched-groups design we may match the groups on the basis of two or more criteria. In the example used above we might have matched the groups on the basis of pretest performance as before, and, in addition, on the basis of near visual acuity. Thus, we start with two groups of subjects, each with the same *average* pretest scores and with the same *average* visual acuity. To evaluate the significance of the difference between the outcome of the trials with A and B, use formula [23] above but with $R_{A(C_1, C_2)}$ replacing r_{CA} and $R_{B(C_1, C_2)}$ replacing r_{CB}. The term $R_{A(C_1, C_2)}$ is the multiple correlation between scores on A, on the one hand, and scores on the two matching criteria, C_1 and C_2, on the other. The $R_{B(C_1, C_2)}$ has a similar interpretation.

In experiments using matched subjects, or in which each subject serves as his own control, the appropriate form of statistical analysis is that illustrated in Tables 15 and 16 in the preceding chapter, or we may use this alternate formula:

$$t = \frac{M_A - M_B}{\sqrt{\dfrac{\sigma^2_A}{N_A} + \dfrac{\sigma^2_B}{N_B} - 2\, r_{AB}\, \dfrac{\sigma_A}{\sqrt{N_A}}\, \dfrac{\sigma_B}{\sqrt{N_B}}}} \qquad [24]$$

Since in these designs $N_A = N_B = N$, we may write

$$t = \frac{\sqrt{N}\, (M_A - M_B)}{\sqrt{\sigma^2_A + \sigma^2_B - 2\, r_{AB}\, \sigma_A\, \sigma_B}} \qquad [25]$$

where N is the number of matching pairs (in the matched-subjects design) or the number of subjects (in the same-subject design), and r_{AB} is the correlation between the scores earned under conditions A and B. The t is evaluated with $N - 2$ degrees of freedom.

Now that we have these formulae in front of us, we can see why matching groups or subjects is better than picking up subjects at random. The denominators of equations [20], [23], and [24] are all expressions for the standard error of the difference between two means. To increase the sensitivity of an experiment, we want to increase t for a given value of $M_A - M_B$. This means, in turn, that we must decrease the standard error. When we use matched groups or matched subjects we do precisely that. The standard error is decreased by virtue of the correlation we have deliberately designed into the experiment.

On what basis should matching be carried out? Formulae [23] and [24] also help us to decide on what basis we should match either groups or subjects. What we should try to do is to match on some variable which will correlate, in a statistical sense, with the variable we plan to measure in the study. That is, we want the r's in formulae [23] and [24] to be as high as possible. Matching never ruins an experiment; it may, however, provide no advantage over a design using independent groups if the variable on which the matching is made does not correlate with the dependent variable being measured in the study, that is, if the r's are equal to zero. For example, if we were doing a study of instrument dial reading, it would be pointless to match subjects on the basis of their hair color, arm reach, or strength of grip, because there is no evidence that any of these factors is related to dial reading. We can do such matching if we want; we would simply fail to gain anything by doing it. Under these conditions, in fact, there will usually be a slight loss in statistical precision because a matched-subjects (or same subjects) design has roughly one-half as many degrees of freedom as a comparable independent-group design. In addition, we will have wasted some time and effort in carrying out this matching, and these are certainly factors of importance.

In a completely new kind of experiment where there is little previous evidence to go on, the experimenter may not know on what basis he should match groups or subjects. In such cases it is usually profitable to arrange a pretest, a small sample of the actual task he will use in the main experiment, simply for the purpose of getting scores on which to match subjects. In their study of dial reading, for example, Weldon and Peterson used such a pretest for matching some of their groups.[112]

Increasing the sensitivity of experiments by increasing N. On page 162 I pointed out that for a given value of $M_A - M_B$ the sensitivity of an experiment could be increased by decreasing the standard error of the difference between the two means. That discussion also showed that one way of decreasing the standard error is by deliberately designing correlation into the experiment. Now we turn to a second way of decreasing the standard error, namely, by increasing the number of subjects tested. The number of subjects, N, appears as a denominator in each standard error term in formulae [20], [23], and [24], but it appears as a square root. This means that, other things equal, you can double the sensitivity of an experiment by quadrupling N. You can triple the sensitivity of an experiment by using nine times as many subjects, or by collecting nine times as many observations.

There are two general implications which follow from these relations. First, one soon reaches a point of diminishing returns in applying this strategy. Imagine that you have performed an experiment with 10 subjects and want to cut the standard error in half, that is, from 100 per cent to 50 per cent. To accomplish this result you must test 30 more subjects, or 40 in all. Now suppose you want to cut the standard error to one-third its original size. This means you must test 90 subjects in all. Note that the addition of 50 subjects $(90 - 40)$ produces a much smaller gain (a reduction in the size of the standard error from 50 per cent to only 33.3 per cent) than did the addition of the first 30. Adding still another 70 subjects (so that the total number now equals 160) produces an even smaller proportionate drop in the standard error—from 33.3 to 25 per cent of its original size. Clearly you soon reach a point where the expense of collecting additional data is no longer worth the potential gain in sensitivity of the experiment.

The second point to notice is that increasing the number of cases could, theoretically at least, make an average difference of any size come out to be statistically significant. The trouble is that with a sufficiently large number of cases you might demonstrate a statistically-significant difference which is practically trivial. Refer again to the discussion of this point in the preceding chapter (pages 144 ff.).

Increasing the sensitivity of experiments by decreasing variability. The third way of increasing the sensitivity of an experiment is by decreasing the inherent variability in the data, that is, by decreasing the σ's in formulae [20], [23], and [24]. In general, there are two ways of doing this: (a) by increasing the precision of the measurements taken in an experiment; and (b) by designing tight controls in the experiment so that irrelevant variables do not contribute significantly to the variability of the data.

Each of these techniques can be illustrated by reference to the data of Table 13 in the preceding chapter. The data there show that the mean reaction time on Stove 1 was 0.56 seconds; that on Stove 3 was 0.66 seconds. Assume that this difference of 0.10 seconds is the size of the true difference between these stoves and that this is the effect we are trying to measure and assess. What would have happened if the experimenter had used the small second hand on his wrist watch to measure the reaction times of the subjects? It does not require very much sophistication to see that this would not have been satisfactory. It would not have been satisfactory because this method of measurement is so inexact that it might easily conceal the effect we are trying to measure. Lack of precision in the measurements shows up in the variability of the data, that is, in increased σ's.

If the experimenter had used a stop watch to record the reaction times, he would have increased the precision of the measurements appreciably. Precision of the measurements could be increased still further (and the σ's reduced correspondingly) by eliminating the experimenter's own reaction time from the measurements. One way of doing this is to have the subject's own hand movement stop an electric time clock, as was actually done in this experiment.

To illustrate the second point mentioned earlier, that is, that lack of control affects the variability of the data, suppose the experimenter had not specified the starting position of the subject's hand preceding each trial. One subject might have started with his hand in his pocket; another with his hand resting lightly on one of the knobs on the front of the apparatus. The time you recorded for the first subject would undoubtedly be too large, that for the second too small because of this uncontrolled factor. This in turn would increase the variability of the data and so increase the σ's. Increased variability can arise from failure to control all sorts of variables in an experiment: the conditions of test; the environment in which tests are conducted; the state of the subject; the instructions given the subject; and even the way in which the experimenter conducts the tests.

Just as in the case of the number of subjects, however, there is a point of diminishing returns here as well. It is all very well to increase the precision of the measurements in an experiment, but it can be overdone. In the stove experiment, for example, we might have substituted a high-precision electronic chronoscope for the electric time clock. There would have been a gain in precision, but the net effect on the outcome of the experiment would probably have been inconsequential. Similarly, we might have controlled what the subjects wore when they reported for the experiment and insisted that they all wear short-sleeve shirts. Or we might have controlled what the subjects had had to eat for breakfast on the day they appeared for the test. However, the gains to be anticipated from controlling these and hundreds of other variables would almost certainly not have been worth the time and trouble. Control is a worthwhile goal, but you have to be realistic and stop somewhere or you will never get an experiment done. Experimenters generally try to control those variables which are likely to have a substantial effect on the variability of the data. Oftentimes they can rely on previous experience, or the results of other experiments, to identify these important variables. Sometimes, however, they can only rely on their own good judgment.

MULTI-VARIABLE EXPERIMENTS

This section will (a) show how a typical multi-variable experiment is constructed from a simple one, and (b) discuss some of the important reasons why human engineers often use multi-variable experiments in preference to the simpler types. Although the example I shall talk about is artificial, it is a meaningful type of human engineering experiment and one for which you can find real counterparts.

A comparison of two warning devices. Imagine that we want to compare two auditory warning devices. Since our interest is in the formal structure of the experimental design, we shall ignore many important and relevant considerations which are discussed later in this chapter and in the next. For the moment focus your attention only on the design itself. Following the principles discussed in the preceding section, we might start by designing the experiment as in Table 20. This is a simple, independent-group experiment. The eight subjects (*A* through *H*) who are to be tested with Warning Device 1 are all different from the eight (*I* through *P*) who are to be tested with Device 2. One way of evaluating the significance of the mean difference obtained with the two devices would be to use the *F*-test (pages 127 ff.) with 1 and 14 degrees of freedom.

The addition of a second variable. One of the relevant variables which influences the effectiveness of an auditory warning device is the background of noise against which it is heard. If we are to obtain a fair comparison of these two warning devices, we need to do something about controlling this factor. In the traditional type of experiment illustrated in Table 20, we would select one kind of noise and hold it constant for all trials with both devices and for all subjects. But what sort of a noise should we pick? Some noises have predominantly high-frequency components, others have relatively flat spectra, while still others have predominantly low-frequency components.

TABLE 20. *An experimental design to compare two auditory warning devices. A through P are different test subjects*

Warning device 1	Warning device 2
A	*I*
B	*J*
C	*K*
D	*L*
E	*M*
F	*N*
G	*O*
H	*P*

More important than this, however, is the strong possibility that the effectiveness of these warning devices will vary with the type of noise you use. It could easily turn out, for example, that Device 1 can be heard much better than 2 when the noise is made up of predominantly high-frequency components, but that 2 can be heard better than 1 in noise which has a preponderance of low-frequency components. In short, you might very well influence the outcome of the experiment one way or the other simply by the choice of the noise background which is held constant.

One way of resolving difficulties of this sort is *to design the tests deliberately to include noise backgrounds of various kinds.* To keep our example simple, let us take only two kinds of noises—the noise of a propeller-driven aircraft and the high-pitched whine of a jet engine. Putting this additional variable into the design does not mean doing a

The Experimental Method 167

TABLE 21. *This is the same experimental design as in Table 20 but with another variable added*

	Warning device 1	Warning device 2
Propeller noise background	A B C D	I J K L
Jet noise background	E F G H	M N O P

bigger experiment. We simply split the two subject groups as shown in Table 21, assigning one half of each group to one kind of noise; the other half to the other kind of noise.

The appropriate form of statistical analysis for such an experiment is illustrated in Table 22. When statisticians discuss the analysis of experiments in the abstract they commonly identify only the sources of variation and the number of degrees of freedom associated with each source. You should understand, of course, that if this were a real ex-

TABLE 22. *Basic form of the analysis of variance appropriate for the experimental design in Table 21*

Source of variation	Degrees of freedom
Between warning devices (W)	1
Between noise backgrounds (N)	1
Interaction: Warning device × noise (W × N)	1
Error (between subjects within groups)	12
Total	15

ample, with real data, we would have additional columns in Table 22 for the sums-of-squares, the estimated variances, and the F-tests. Table 22 is simply an abridged version of Table 14, 16, or 18. Although this analysis is a little different from any in Chapter 4, if you understood the basic principles of analysis discussed there, you should have no difficulty in seeing how you would go about computing the sums-of-squares for data in this form. Note that the interaction term would be

TABLE 23. *This is the same experimental design as in Table 21 but with another variable added*

		Warning device 1	Warning device 2
Propeller noise background	90 db	A B	I J
	120 db	C D	K L
Jet noise background	90 db	E F	M N
	120 db	G H	O P

computed with the help of a 2 × 2 sub-table in exactly the same way as was illustrated in Table 19.

Adding a third variable. Another factor we should control before running these tests is the intensity of the noise background. Here again we face exactly the same kind of problem we had when we were trying to settle on the kind of noise to use in our tests. The *amount* of noise will certainly influence the audibility of these two warning devices, and, moreover, it might have a *differential effect* on the audibility of the two signals. The words *differential effect* imply another interaction, of course.

Once more we can, if we wish, deliberately design this additional variable into our tests. Again to keep our example small, let's take only

The Experimental Method 169

TABLE 24. *Another way of diagraming the experimental design in Table 23*

Warning device		1				2		
Type of noise	Propeller		Jet		Propeller		Jet	
Intensity of noise in *db*	90	120	90	120	90	120	90	120
Subjects	A	C	E	G	I	K	M	O
	B	D	F	H	J	L	N	P

two intensities—90 and 120 *db*. The four groups in Table 21 are each split in two. One-half of each group is assigned to the 90 *db* intensity level; the other half to the 120 *db* intensity. The result is shown in Table 23. It is important to point out that there are several alternative, but equivalent, ways of diagraming such experimental designs. Table 24, for example, shows one of them.

The statistical analysis for a set of data like that in Table 23 or 24 is shown in Table 25. Notice now that we have three double interactions, $W \times N$, $W \times I$, and $N \times I$. These are each computed as for Table 22. Note further that we have a new kind of interaction, a triple

TABLE 25. *Basic form of the analysis of variance appropriate for the experimental design in Tables 23 or 24*

Source of variation	Degrees of freedom
Between warning devices (W)	1
Between noise backgrounds (N)	1
Between intensities of noise (I)	1
Interactions: $W \times N$	1
$W \times I$	1
$N \times I$	1
$W \times N \times I$	1
Error (between subjects within groups)	8
Total	15

interaction, $W \times N \times I$. To compute the sum-of-squares for the triple interaction we proceed in exactly the same way as we did in computing the double interactions. First we prepare a sub-table (Table 26). Eight numbers are entered into this table. They are the sums of the scores, or measurements, for the two subjects who were

TABLE 26. *The sub-table used for computing the triple interaction $W \times N \times I$, for Table 25*

		Warning device 1	Warning device 2
Propeller noise	90 *db*	$A + B$	$I + J$
	120 *db*	$C + D$	$K + L$
Jet noise	90 *db*	$E + F$	$M + N$
	120 *db*	$G + H$	$O + P$

tested under identical conditions, i.e., $A + B$, $C + D$, $E + F$, etc. The total sum-of-squares for Table 26 is then:

$$\frac{(A + B)^2 + (C + D)^2 + (E + F)^2 + \cdots + (O + P)^2}{2}$$

$$- \frac{(A + B + C + D + \cdots + O + P)^2}{16}$$

From this total sum-of-squares subtract the following six terms:

1. the sum-of-squares between warning devices, W,
2. the sum-of-squares between noise backgrounds, N,
3. the sum-of-squares between intensities of noise, I,
4. the sum-of-squares for the interaction, $W \times N$,
5. the sum-of-squares for the interaction, $W \times I$, and
6. the sum-of-squares for the interaction, $N \times I$.

What is left is the sum-of-squares for the triple interaction, $W \times N \times I$.

On the meaning of a triple interaction. A significant triple interaction means that the double interactions change for different values of the third variable. There are usually a large number of ways in which a significant interaction can occur with any particular set of numbers. The numbers in Table 27 show one way. These are arbitrary numbers which, for purposes of this example, we may assume are positively correlated with the audibility of the signal—the higher the number the more audible the signal.

The upper set of numbers represents the data obtained at a noise level of 90 *db*. We see there an interaction between warning devices

TABLE 27.   *These values have been selected to illustrate a triple interaction*

	Device	
90 *db*	1	2
Propeller noise	95	60
Jet noise	60	95

	Device	
120 *db*	1	2
Propeller noise	75	40
Jet noise	30	30

and noise background, $W \times N$. Device 1 is more audible than 2 in propeller noise, but 2 is more audible than 1 in jet noise.

The lower set of numbers represents the data obtained at the noise level of 120 *db*. These numbers show another double interaction. Device 1 is still better than 2 in propeller noise, but notice that at this high noise level neither 1 nor 2 is effective in jet noise.

If these two double interactions showed the same *pattern* at the two intensity levels, there would be no triple interaction in these data. That is, if the two 30's in the bottom of the second square were replaced by a 40 and 75, in that order, the triple interaction would be zero. As they stand, however, the data in Table 27 show that the double interaction between W and N changes when we go from one intensity level to another. It is this situation that produces a large triple interaction.

There are two other identical ways of illustrating the same triple interaction. Table 28 shows one of these. It is expressed in terms of the double interactions between W and I at different values of N. The third way, which you can set up by yourself as an exercise, is in terms of the double interactions between N and I for different W's.

TABLE 28.   *These are the same values as appear in Table 27 arranged somewhat differently*

Propeller noise	Device	
	1	2
90 *db*	95	60
120 *db*	75	40

Jet noise	Device	
	1	2
90 *db*	60	95
120 *db*	30	30

The advantages of multi-variable designs. Now that we have constructed a multi-variable design and worked through its analysis and interpretation, let us see what advantages such a design has. Look first at Table 20. We have eight observations (A through H) on Device 1, and these can be compared with eight observations (I through P) on 2. If we have planned our experiment well, the only *systematic*, or deliberately planned, difference between the two groups of measurements is the type of warning device used. Type of noise and intensity of noise have been controlled and are the same for all observations. There will be differences between the individual subjects, of course, but if we have chosen them correctly the differences should be random and should not favor either warning device.

Now look at Table 23. How many observations are available for comparing 1 with 2? Clearly we can compare A and B against I and J. All four were tested with the same noise and with the same intensity of noise. The only systematic difference between these two pairs of observations is the type of warning device used. In addition, we can compare C and D with K and L. All four of these were also tested with the same noise and with the same intensity of noise. The only systematic difference between these two pairs of observations is the type of warning device used. Exactly the same kind of argument applies to E, F and M, N; and to G, H and O, P.

To sum up, in comparing 1 with 2 by means of the design illustrated in Table 23 we can compare A, B, C, D, E, F, G, and H against I, J, K, L, M, N, O, and P. This is exactly as many comparisons as we had available in Table 20.

But the same set of 16 observations in Table 23 allows us to make another kind of comparison which we could not make from the data in Table 20. Look at items A, B, E and F. These four observations were made with the same warning device and with the same intensity of noise. The only systematic difference between them is that A and B were tested in prop noise whereas E and F were tested in jet noise. If you continue in this way you will see that the only systematic difference between A, B, C, D, I, J, K, L and E, F, G, H, M, N, O, P is the kind of noise used. This amounts to another complete experiment of the type shown in Table 20, but this second major comparison is obtained with the same sixteen observations we started with.

But the story is still not finished. The only systematic difference between A, B, E, F, I, J, M, N and C, D, G, H, K, L, O, P is the intensity of the noise used. This amounts to another complete experiment of the type shown in Table 20.

Now let's summarize what we have found out. The same sixteen observations in Table 23 allow us to make comparisons between three major pairs of conditions. The only thing we had to do was to take the sixteen observations in different combinations for the different comparisons. If we had relied on the type of design shown in Table 20, we would have had to do *three* separate experiments to get information on as many variables as were tested by the *single* experiment in Table 23. This is one of the features of multi-variable experiments which makes them so much more efficient than single-variable ones.

Compensating for the loss of degrees of freedom. In going from the design illustrated in Table 20 to that in Table 23 there is a slight loss in precision for any one comparison because of the reduction in the number of degrees of freedom available. The analysis of variance for the data in Table 20 (page 166) has 14 degrees of freedom available in the denominator of the F-test for evaluating the significance of the difference between the two warning devices. Table 25, however, shows that there are only 8 degrees of freedom available for the same evalua-

TABLE 29.   *This is the same experimental design as in Table 23 except that an additional eight subjects have been added to increase the precision of the statistical tests*

		Warning device 1	Warning device 2
Propeller noise background	90 db	A B C	M N O
	120 db	D E F	P Q R
Jet noise background	90 db	G H I	S T U
	120 db	J K L	V W X

tion when we use the design illustrated in Table 23. In general, such a reduction in number of degrees of freedom means a slight loss of precision. An F must have a value of 4.60 to be significant at the 5 per cent level with 1 and 14 degrees of freedom. With 1 and 8 degrees of freedom, however, the F must be at least 5.32 to reach the same level of significance.

This loss in sensitivity is very easily compensated by adding another group of eight subjects as shown in Table 29. The basic form of analysis for the data in Table 29 is exactly like that in Table 25 except that the total number of degrees of freedom is 23 and the error term will have 16 degrees of freedom. A comparison of Table 29 with 20 shows that the former design gives us three major comparisons instead of one, with a slight gain in precision (because we have 16 instead of 14 degrees of freedom in the error term), for only a 50 per cent increase in the number of measurements that have to be collected.

Multi-variable experiments give us interactions. By now it should be apparent that another important characteristic of multi-variable experiments is that they provide us with interactions. There is no way you can extract an interaction from a set of data like those shown in Table 20. If you want to know about the interplay of different variables in determining experimental outcomes—as you often do in human engineering work—a multiple variable experiment is the kind to use.

Generality of the outcomes. The last important feature of multivariable experiments is closely tied up with the characteristics discussed above. This is that multi-variable experiments give us greater confidence in the generality of our findings. Suppose that we had conducted an experiment of the type shown in Table 20 and that Device 1 turned out to be better than 2 at some appropriate level of significance. What could we conclude? We would know that Device 1 is better than 2 with *one* particular kind of noise (because this factor was, or should have been, controlled), and at *one* particular intensity of noise.

Contrast this with what we could conclude if the statistical analysis in Table 23 showed that Device 1 was better than 2. Now we would know that 1 is better than 2 for two kinds of noise and for two different intensities of noise. When we have deliberately designed a number of different variables into our experiment, we gain generality because we know that when we get a significant effect, this effect holds for a number of different situations. This gain in generality is another powerful reason for using multi-variable experiments.

Extensions of this design. The structure of this hypothetical experiment was kept simple so that it would be easy to talk about here. You should understand, however, that the variables could have been extended to include more values than two. For example, we could very easily have used three or four different kinds of noises, instead of only two. We could have used four or five different intensities of noise, instead of two. And, of course, we could have included still other variables, for example, different working conditions. We could run tests under conditions when the experimental subjects have nothing else to do except listen for the warning signal (we might call this a nondistraction work situation), and compare this with results obtained when the subjects were occupied with a difficult task which was irrelevant to the listening job (we might call this a distraction condition). Once you understand the basic principles, it is a simple matter to extend an experimental design to include more variables or conditions.

Summary remarks. This hypothetical study of warning devices is an illustration of a *factorial* design. Factorial experimental designs make up one major class of multi-variable experiments and constitute one of the most important basic designs you will need in human engineering work. The distinguishing characteristic of a factorial design is that each combination of independent variables is tested with a different group of subjects. In Table 23 these groups are each made up of two subjects; in Table 29 they are each made up of three subjects. Thus the factorial design is an extension of the simple *independent-groups* type of experiment discussed on pages 157 to 159.

The factorial design is particularly suited to experiments in which performance under one set of experimental conditions might have an important effect on performance under other experimental conditions. Another way of saying it is that problems of progressive changes in performance—such as result from learning, boredom, or fatigue—are generally minimized with factorial designs. One of the principal disadvantages of the factorial design is that it requires a larger number of experimental subjects than are needed for some other types of design.

Two other important forms of experimental design are (a) the *treatments* × *subjects* design (to be illustrated on pages 179 to 192), and (b) the *Latin square* design (discussed on pages 192 to 198).

On the usefulness of simple comparisons. This heavy emphasis on multi-variable experiments should not lead you to believe that the simple types of comparisons discussed earlier (pages 157 to 165) have no usefulness in human engineering work. On the contrary, they are very useful for getting many practical answers efficiently. However, you should be alert to the possibilities of multi-variable designs for increasing the generality of your experiments.

DESCRIPTION OF SOME
TYPICAL EXPERIMENTS

The discussion so far in this chapter has been concerned with general principles governing the design of human engineering experiments.

Now we shall turn to some case studies to see how these general principles have been put to use in some real experiments. Our interest will be in how the experimenters defined their problems, established controls, and designed their experiments.

MEASURING DARK ADAPTATION

You have undoubtedly observed that if you go into a theater after leaving a brightly lighted street, at first you cannot see the difference between an empty seat and a fat lady. Gradually as you sit in the theater, in an empty seat you may have found by feel, your eyes become more and more adjusted to the darkness until finally, after a half hour or so, you can see quite well. Many people have observed this phenomenon in a general sort of way, but scientists have been interested in formulating a more precise description of this function. How exactly have they proceeded to obtain such a statement?

Formulating the problem. First of all, the scientist needs to formulate the problem carefully. In this case, he is interested in the relationship between the sensitivity of the eye and the length of time the eye has been in the dark. Length of time in the dark is easy enough to measure, but the "sensitivity of the eye" requires a little more thought. There are various measurements one might make under the heading of sensitivity measurements, and the decision about which to use is a matter the experimenter has to settle before he starts. For example, he might measure the faintest amount of light the eye can see; the smallest set of letters or numbers the eye can see at some low level of illumination (a visual acuity type of measurement); or the smallest brightness contrast the eye can see. Any of these would make an interesting and meaningful experiment. Choosing from among them requires that the experimenter decide just exactly what it is he wants to know.

Let us suppose that our experimenter decides to measure the faintest light the eye can see, because he is interested in the practical problem of seeing signal lights on a dark night.

Holding factors constant. Having decided on the two variables which are to be studied systematically, the experimenter will hold everything else constant. What does that mean? It means that he will do his experiment in a completely dark room. Extraneous sources of light, of course, would mess up his results. Next, the size of the test light must be held constant. The same color of test light must be used all the time, because we know that the color of the test light makes a difference. Then, too, we know that if one has just been looking at the sun, his eyes are blinded for quite a long time; whereas if he steps into a dark room from a dimly lighted room he can see fairly well. So, obviously the amount of previous exposure to light must be controlled. We must make some attempt to control a lot of other factors which concern the individual. The test light should stimulate the same part of the eye,

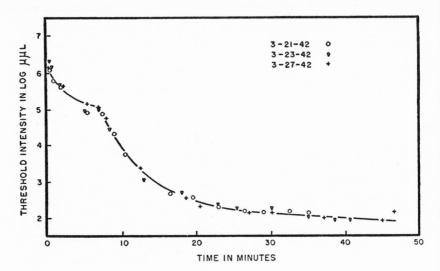

FIG. 27. Dark-adaptation measurements made on the author's eye on three separate days. (After Chapanis.[20])

since different parts of the eye are not equally sensitive to light. The subjects should be well rested, not fatigued, and they should be normal observers with respect to their visual apparatus. To be sure about this last point, he might even have his observers checked by an ophthalmologist. Finally, to get some indication of the random error in this type of experiment, the experimenter might repeat the entire series of observations on each of several different days.

Having thus arranged and controlled all of the various conditions— the completely dark room, the size of the light to be tested, the color of the light, the previous exposure of the subject, and so on—the experimenter can now proceed to investigate the nature of this process known as dark adaptation. He does this by measuring for each subject the amount of light that is just barely visible at various times after going into the dark room. The results of such an experiment yield a curve such as is shown in Figure 27. These particular data were obtained with a violet test light of known spectral composition, 3° in size, stimulating the right eye 7° to the right of the fovea (the central region of most acute vision). Measurements were begun after initial preadaptation for 3 minutes to a light of 1,500 millilamberts.

We see that the process of dark adaptation is really two processes. The sensitivity of the eye increases fairly rapidly during the first few minutes, it then levels off, and in about ten minutes it suddenly shows another very sharp increase in sensitivity. Notice also that although the data for the separate days show some variability, the variability is sufficiently small so that we can draw the over-all function with some confidence. This kind of experiment has been repeated many times, in many laboratories and with all sorts of variations of the conditions which were held constant in the experiment shown in Figure 27. We know now that curves like those in Figure 27 accurately represent the increase in sensitivity of the eye as a function of time in the dark. This is what is meant by a scientific generalization, a functional relationship. This kind of information is extremely important in furnishing us with the basic knowledge necessary for the solution of many problems concerned with seeing at night.

Let us stop a minute and underline what we said about the essential method of science being that of controlled observation. The scientist has to control his conditions in order to be sure of what he is getting. If our test at ten minutes was made with a pinpoint of light and at twenty minutes with a large lamp, our results would be, to put it mildly, contaminated by this extra variable. Obviously, the size of light is an important variable which can be, and has been, studied in its own right. But in order to get meaningful data we must vary this factor systematically if we want to discover its precise contribution to the total effect.

AN EXPERIMENT ON RADAR

For my second case study I shall use a study by Gebhard on rate-aiding equipment for the *VF* radar.[58] As such studies go, this is a simple one, but the design and results are tidy. It is well-suited to our purposes because a small experiment like this one can point up some basic principles of experimental design without complicating the story with an excessive number of details.

Background of the problem. Before we can get to the experiment itself, we need to say a few words by way of introduction. A radar of the type used in this study typically has an antenna which continuously rotates, searching the sky. Geared to the antenna is an electronic beam which sweeps across a cathode-ray tube (the *B*-scope on this radar). Every time the antenna receives an echo from a target, a bright spot, or target "pip," shows up on the *B*-scope (provided that the target is within certain limits of bearing and range). The radar operator has two controls—a bearing knob and a range knob—by means of which he can move the target pip to the center of the *B*-scope. When the pip is centered on the scope, bearing and range dials tell him the bearing (direction) and range (distance) of the target.

Since the antenna rotates only a few times a minute, information about the location of the target is available only once every few seconds. In the Gebhard study, the antenna rotation rate was 8 rpm (revolutions per minute) so that the target pip was renewed once every 7.5 seconds. If the target is moving rapidly, the target pip will appear in a new place on every sweep and the displacement from one sweep to the next may be substantial. In tracking rapidly moving targets, experienced radar operators frequently try to anticipate where the target will next appear, that is, they crank the bearing and range controls to try to bring the target to the center of the scope on the next sweep.

Anticipating a rapidly moving target is difficult to do accurately. First, the operator must try to get some idea of the rate of target movement on the basis of the successive positions of the target during two or more sweeps. Having formed some idea of the rate of target movement, the operator must then turn the two controls simultaneously at varying and different rates, sometimes in the same direction, and sometimes in opposite directions. The rates and directions of turning depend, of course, on the speed of the target and the direction in which it is moving. A further complication arises from the fact that targets often fade, or disappear, from the scope and cannot be picked up for one or more sweeps. We need not go into the electronic reasons for "fading"—we merely need to know that it happens.

To help the operator in this complicated tracking task, the manufacturer devised a variable-rate control to be attached to the two cranks of the radar. With this device, the operator sets in *rates of movement*, and, as long as the target does not change its direction or speed, the mechanical aid will anticipate correctly where the target will be on the next sweep. Of course, there are still some human problems with this device. First, it will *track* accurately only if the operator can set in correct rates of movement. If he does *not*, he has an opportunity to correct the rate of movement on the next sweep by noticing how far the pip misses the center of the scope. In addition, the operator must make corrective adjustments every time the target changes its speed or direction.

Purpose of the Gebhard study. With this background information we can now state the purpose of the study. It was to compare the accuracy with which moving targets could be tracked through fades with and without the rate-aiding attachment.

Design of the experiment. Now let us turn to the design of this experiment, examine it minutely, and discuss some of the controls which were used. The basic design is in Table 30. Statisticians sometimes call this a *treatments* × *subjects* design because each subject is tested with every combination of treatments (or variables as I have been referring to them). Contrast this with the design in Table 23 where each subject is tested with only one combination of treatments (or variables). Note, incidentally, that a *treatments* × *subjects* design is an extension of the simpler type of experiment discussed earlier in which each subject serves as his own control (See page 160).

The first variable is the method of tracking. This, of course, is the primary one the experimenter was interested in. Trials listed under the heading of *Aided method* in Table 30 are those in which operators made use of the rate-aiding attachment; trials listed under the heading of *Manual method* are those in which operators did not. To rule out differences which might exist between two different radars, the same radar was used throughout these tests. The rate-aiding device was installed or removed as required by the test schedule.

To be sure that the two methods were compared fairly, operators were given extensive practice on both methods of tracking. In addition, comparisons had been made (in a preliminary experiment) between the errors made in the first three of a series of scans and the last three of a series of scans. Since there was no difference between the two sets of data, it looked as though the subjects were thoroughly practiced in both techniques.

To insure further that the two methods were given a fair test, a trial consisted of continuous tracking until the operator was satisfied that he was accurately following the target. Only then did he signal the experimenter that he was ready for the test to begin. This procedure helped to control for the readiness of the operator. If some trials were started before the operator had "gotten into the groove," whereas

TABLE 30. *Design of an experiment to test whether a mechanical rate-aiding attachment helps radar operators track targets through fades of the target on the B-scope of the VF radar. The numerical entries show the order in which tests were made. The letter entries correspond to the four different target courses (after Gebhard[58])*

		Aided method		Manual method	
		2-sweep fade	4-sweep fade	2-sweep fade	4-sweep fade
Operator A	Trial 1	2 a	10 d	4 b	8 c
	Trial 2	3 c	12 b	7 d	13 a
Operator B	Trial 1	6 b	1 c	9 a	5 d
	Trial 2	15 d	11 a	14 c	16 b

others were started after an ample ready period, this would have introduced an uncontrolled variable into the study.

The second variable in this experiment was the length of the fade. It seemed likely that a target which faded for a longer period might be harder to track than one which faded for a shorter time. In this experiment, two different lengths of fade were used: a 2-sweep fade; and a 4-sweep fade. The location of these headings in Table 30 shows that both kinds of fade were tested with both tracking methods.

The third variable in this experiment is the operators. Although an experiment like this could be run with a single subject, the experimenter has more confidence in his findings if he gets the same results on more than one person. In this experiment, there were two operators and, as we shall see when we discuss the findings, both gave substantially the same results. This agreement between operators makes it less likely that the findings were due to some peculiar quirk or characteristic of a particular operator. One other kind of control which is important in this connection is the past experience of the subject. We can never get perfect control in this respect, but we can at least avoid some of the obvious pitfalls. It would not do, for example, to use a trained radarman as one subject, and a naive high-school girl who had never heard of radar, bearings, or ranges as the other subject. In the present case, both operators were trained Navy radarmen. You will also remember from the discussion above that they had been given extensive practice on both tracking methods to equate their experience still further.

You will notice from Table 30 that each subject was tested *twice* with each combination of tracking method and length of fade to provide one estimate of experimental error which we need.

The numerical entries in Table 30 show that the tests were not given in any regular order. For example, on the first test, Operator *B* was tested with the aided method and a 4-sweep fade. On the second test, Operator *A* was tested with the aided method and a 2-sweep fade, and so on. These numbers were deliberately picked out of a table of random numbers in order to avoid bias in the data.[48, 93] As mentioned earlier in this chapter, randomizing trials is one way an experimenter can guard against systematic effects which he might not have thought of, or might not be able to control. What are some of these biasing effects? Anything which changes the behavior of the subjects (learning, fatigue, boredom) or the behavior of the equipment (changes in calibration, wear and tear) throughout the course of a series of trials could bias the data.

For example, let us suppose that the calibration of the radar changed throughout this series of tests. We start with the radar well calibrated, but it starts to drift during the course of the tests. If we had run all trials with the aided method first and all the trials with the manual method second, the manual method would undoubtedly show up with the larger errors but it might be due, not to an inherent difference between the two methods, but only to the fact that the manual method was tested when the instrument was less stable. In short, *randomizing* the order of presentation of trials is one way of avoiding the systematic bias which might contaminate the data because of changes which occur in equipment, subjects, or the experimenter during the course of an experiment.

The targets. There is still another part of the study we have not considered. This concerns the targets for test. Although it would have been possible to use live targets—real aircraft flying various courses around the radar—for a study like this there are two practical objections to this kind of target. The first is expense. The second comes back to the matter of control. It is extremely difficult to achieve adequate control over the movements of live targets. It is almost impossible for pilots to fly identical courses, at identical speeds and altitudes, enough times to satisfy the requirements of an experiment. Unless we have some control over this variable, we cannot be sure that we are giving our tracking methods a fair trial.

In this experiment a target generating system produced artificial targets whose movements were controlled by a set of bearing and range cams. The cams were cut to simulate attack courses and there were four, indicated by the letters *a*, *b*, *c*, and *d* in Table 30, used in this study. The courses changed both in bearing and range.

The target courses were assigned for test in a counterbalanced order, and there is again a very good reason for this. Notice first that on each trial each operator uses each of the four courses. Then note that he uses the four target courses in a different order on the two trials. The reason for this is to make it less likely that the subject could recognize or remember a particular course. Next, notice that in any column of Table 30 each target course appears once and only once. The reason for this is to avoid the possibility of biasing the methods under test by assigning to them target courses which were uniformly more difficult or easier than average. In this experiment, for example, target course *d* had the most rapid changes of bearing. If this course had been assigned exclusively to the trials with the manual method, it might make the manual method come out worse only because it was such a difficult course to follow. By assigning courses to the methods in this systematic way (not randomized, in this case) we can be sure that each method is tested with both easy and difficult courses.

The measurements. To get precise measurements of the kind needed for analysis, a 16-millimeter motion-picture camera was mounted over the *B*-scope. This camera could be started and stopped at any time. Measurements of bearing and range errors could then be made from the photographic record of where the target was with respect to reference co-ordinates.

TABLE 31. Range errors (in yards) made by two operators in tracking a moving target through fades on the B-Scope of the VF Radar. A plus error means that the radar setting is too big; a minus error that the radar setting is too small (after Gebhard[58])

	Aided method								Manual method							
	2-sweep fade				4-sweep fade				2-sweep fade				4-sweep fade			
	P_1	P_2	P_3	P_4	P_1	P_2	P_3	P_4	P_1	P_2	P_3	P_4	P_1	P_2	P_3	P_4
Operator A Trial 1	10	10	10	15	-20	-55	-60	-10	-10	-10	0	40	-40	-190	-15	-50
Operator A Trial 2	25	130	100	30	20	40	-10	30	30	-155	-90	40	15	-305	60	100
Operator B Trial 1	30	0	-15	-30	0	-15	-30	-55	-55	-90	-65	0	-15	170	80	-100
Operator B Trial 2	-30	-100	15	25	10	-45	25	55	10	-95	45	-40	-30	100	0	-10

P_1 = the position of the target on the last sweep *before* the fade.
P_2 = the position of the target on the first sweep *after* the fade.
P_3 = the position of the target on the second sweep *after* the fade.
P_4 = the position of the target on the third sweep *after* the fade.

There is one other point of procedure which is important. There were actually four separate measurements made for each cell of Table 30. When the radar operator was satisfied that he was ready, the experimenter photographed the target and the reference co-ordinates, the pip faded for either two or four sweeps, and the subject tried to anticipate the next appearance of the target by appropriate movements of

TABLE 32. *The data of Table 31 are summarized here to show the average range errors obtained under various test conditions. In computing these averages the direction of the error (that is, whether it is plus or minus) has been ignored (after Gebhard[58])*

Condition	Average error in yards
Aided method	33
Manual method	64
2-sweep fade	42
4-sweep fade	55
Position 1	22
Position 2	94
Position 3	39
Position 4	39
Operator A	54
Operator B	43
Average Error (all conditions)	49

his controls. When the target pip appeared again, the camera photographed the following three sweeps. Thus, the data consist of errors made on the sweep immediately before the fade and on the three sweeps following the fade.

Results. The results of this experiment are shown in Table 31. There were two separate kinds of data which came out of this experiment—bearing and range errors—but we shall confine our discussion to the range errors. Since it is difficult to make much sense of the raw data in

Table 31, they are summarized in Table 32 in a way which is a little easier to understand.

The first comparison is for the two different methods of tracking. There we note that the average error of tracking for the aided method is about half that of the manual method. Incidentally, we might point out that the value of 33 is obtained by adding the 32 measurements under *Aided method* in Table 31 and dividing by 32. The value of 64 is the result of a similar computation.

The second point of interest concerns the length of the fade. Although the 4-sweep fade gives larger errors than the 2-sweep fade, the difference is slight. To get the numbers 42 and 55, we average the data of Table 31 in a slightly different way. This time we take the 16 measurements under *2-sweep fade, Aided method*; and add them to the 16 measurements under *2-sweep fade, Manual method*. This sum divided by 32 gives us the value 42. The number 55 comes from averaging the 32 measurements under *4-sweep fade; Aided method* and *4-sweep fade; Manual method*.

The third comparison has to do with the position of the target with respect to the fade. The average error is smallest for position 1. This is understandable because this is the error on the last sweep before the fade. The operator has been tracking, is "in the groove," and there is reason to expect that he is doing about the best he can do at this point. The error is very high for position 2, and then drops for positions 3 and 4. Again, this is reasonable when you look at the task. Position 2 is the first sweep after the fade. Difficulties in moving the controls during the fade (with the manual method) should show up in a large error when the target first appears after the fade. Similarly, if the target changed direction or speed during the fade this will show up as an error in the aided method immediately after the fade.

Finally, we may note that there is a small difference between the two operators. Operator *B*, in general, was able to track with smaller errors than Operator *A*.

Statistical analysis of the data. Table 32 tells us that there are a number of differences in our data and that some are big and some are small. Which ones can we trust? The statistical analysis of these data appears in Table 33. Although this analysis is a bit more complicated than anything discussed in Chapter 4, the basic procedure is exactly as outlined there. As an exercise, start with the data in Table 31 and see if you can come out with the same answers as appear in Table 33.

In the factorial design analyzed earlier in this chapter there was only

TABLE 33. *Analysis of variance of the range error data in Table 31*

Source of variation	Degrees of freedom	Sum of squares	Estimate of variance	F-ratio
Between methods (M)	1	15,625.00	15,625.00	15,625.00/351.56 = 44.44*
Between fades (F)	1	2,626.56	2,626.56	2,626.56/756.25 = 3.47
Between positions (P)	3	47,867.19	15,955.73	15,955.73/1,135.41 = 14.05**
Between operators (O)	1	1,806.25	1,806.25	
Interactions:				
M × F	1	5,625.00	5,625.00	5,625.00/2,139.07 = 2.63
M × P	3	18,562.50	6,187.50	6,187.50/1,076.56 = 5.75*
M × O	1	351.56	351.56	
F × P	3	6,817.19	2,272.39	2,272.39/1,243.75 = 1.83
F × O	1	756.25	756.25	
P × O	3	3,406.25	1,135.41	
M × F × P	3	10,662.50	3,554.16	3,554.16/805.72 = 4.41
M × F × O	1	2,139.07	2,139.07	
M × P × O	3	3,229.69	1,076.56	
F × P × O	3	3,731.25	1,243.75	
M × F × P × O	3	2,417.18	805.72	805.72/1,698.43 = 0.47
Within class	32	54,350.00	1,698.43	
Total	63	179,973.44		

* Significant at the 10% level.
** Significant at the 5% level.

one error term for computing all *F*-ratios (as, for example, in Table 25). That error term was made up of the sum of the squared deviations between the scores of individuals tested under identical experimental conditions. Illustrations of this procedure were worked out in connection with Tables 11 and 13 in Chapter 4.

The pattern of the Gebhard experiment, however, resembles the subjects-times-experimental conditions design illustrated in Table 15 of Chapter 4. There you will recall that the error term was an interaction of subjects-times-experimental conditions. To apply the same rule here means that we must use different error terms for testing different experimental conditions.

Consider first the difference between methods. Preparing an appropriate sub-table of the data gives us:

	Aided method	Manual method	Mean
Operator *A*	36	72	54
Operator *B*	30	57	43
Mean	33	64	

All entries in this table have been rounded off to the nearest whole number. The number 36 is the mean of the 16 scores made by operator *A* when he used the aided method. The three other basic entries were obtained in a similar fashion.

When the data are arranged in this way, the appropriate form of analysis becomes obvious. We can compute a total sum-of-squares with 3 degrees of freedom, a sum-of-squares due to the methods with one degree of freedom, a sum-of-squares due to operators with one degree of freedom, and a residual sum-of-squares also with one degree of freedom. The residual in this case is the interaction of operators-times-methods, $M \times O$.

Now consider the difference between fades. Again we make up a sub-table, but in the following form:

	2-sweep fade	4-sweep fade	Mean
Operator *A*	44	64	54
Operator *B*	40	46	43
Mean	42	55	

A study of this sub-table will show that the appropriate error term for evaluating the difference between fades is the fades-times-operators interaction, $F \times O$.

Table 33 shows all of the F-ratios you might want to compute with their appropriate error terms. Incidentally, experimenters are almost never interested in the differences between subjects. People differ significantly. We know this from thousands of experiments. Finding out that subjects differ in still another experiment is, therefore, of trivial interest except possibly to reassure us that we selected representative subjects.

Results of the Gebhard experiment. Since this was a small, exploratory study, we should probably be a little more lenient than usual in interpreting the findings. Those F-ratios which are significant at the 10 per cent level or better are identified in Table 33. We see that three meet our criterion. First is the difference between methods. Second is the difference between positions. Third is the interaction between methods and positions of the target $(M \times P)$.

The interaction says that the difference between the two methods is not the same at the different positions. To understand what is happening here, we need to summarize the data into finer categories than we used in Table 32. Figure 28 shows the average errors made with the aided and the manual methods, but the data are shown separately for the four positions. Now we can easily see the source of the interaction. For positions 1, 3, and 4, the manual method gives errors which are about 40 per cent greater than the errors with the aided method. However, when we look at the data for position 2 we find that the manual method gives errors which are over two and one-half times as large as the average error with the aided method. Although the aided method is uniformly better than the manual method, its superiority is most pronounced for targets at position 2, that is, on the first sweep after a fade.

Some summary observations on the Gebhard experiment. By way of general summary, there are a couple of points we ought to make clear about the Gebhard experiment. The first is that ordinarily we would not plan as small an experiment as this for most types of human engineering studies. For example, we would feel more comfortable if there had been more subjects, and perhaps more targets than were used in his study. It is a rare event when we get such neat data from so small a study. Considering the levels of significance we were willing to

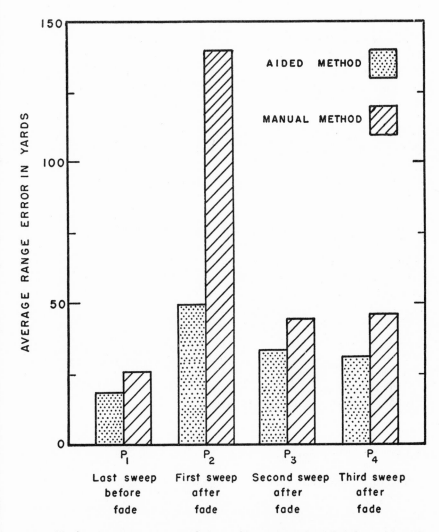

FIG. 28. Average range errors made in tracking targets through fades on the *VF* radar. These data were obtained from Table 31. (After Gebhard.[58])

accept in Table 33, the chief function of this study was to show us that it would probably be worthwhile undertaking a larger experiment to explore these variables more thoroughly. That would be the next step.

The second general observation which needs emphasis is that the conclusiveness of these findings constitutes in part a testimonial to the meticulousness of the experimenter. This experiment did not just happen. It was *planned*—planned down to the minutest detail. The choice of subjects, their training, the selection of targets, the choice of variables, the order in which tests were administered—all these details had to be thought out and anticipated before the first trial was run. Only by careful planning of this sort do we end up with data which mean what they seem to mean.

Some summary observations on the treatments × *subjects design.* The chief advantages of this design are efficiency and economy. It is basically an extension of the simple type of experiment in which each subject serves as his own control (See page 160) and has for this reason the efficiency characteristic of such plans. This type of design also needs fewer subjects than would be needed for a comparable factorial experiment—a feature of some importance in human engineering work, especially if subjects need to be extensively trained before they can participate in the experiment.

The chief disadvantages of this design are that (a) each subject may have to be available for an extended series of tests; (b) randomization or counterbalancing of trials to guard against bias from progressive changes in performance is critical; and (c) it cannot be used for certain experiments in which participation in some of the experimental conditions may interfere with performance in others.

A LATIN-SQUARE DESIGN

The final case study involves an experimental design quite different from any discussed so far. Its special feature is that it makes use of an almost perfect scheme for counterbalancing trials and so produces a very efficient way of testing simultaneously the effects of three variables. In this particular instance it was used to measure the effects of learning without having the learning contaminate the two other main effects being tested.

Shown in Table 34 is the design used in an experiment to find out whether tilting a toll-operator's keyset would have any effect on the operator's keying time. We need not know very much about the apparatus here beyond the fact that it has ten buttons, that it sits on the right-hand side of the toll-operator's work space, and that normally it is flat.

There are three variables in this experiment: (1) subject; (2) inclination of the keyset; and (3) the day on which trials were made. Eight subjects were tested on the equipment, and they are represented by the letters *A*, *B*, *C*, *D*, etc., in the left-hand column in this table. Across the top of the table are the inclinations of the keyset which were

TABLE 34. *Latin-square experimental design used in testing the effect of tilting the toll-operator's keyset. The Arabic numerals are the days on which each operator was tested with each inclination*

Subjects	Inclination of keyset							
	0°	5°	10°	15°	20°	25°	30°	40°
A	2	7	4	8	5	1	3	6
B	1	5	2	4	6	7	8	3
C	3	6	5	2	8	4	7	1
D	8	2	3	7	1	6	4	5
E	7	8	1	3	4	5	6	2
F	5	1	8	6	7	3	2	4
G	6	4	7	1	3	2	5	8
H	4	3	6	5	2	8	1	7

tested. The numerical entries in the table represent the order in which the various combinations of subjects and inclinations were tested. For example, on the first day, operator A was tested with the keyset inclined to 25°; on the second day, she was tested with the keyset inclined to 0°; and so on. You will note that these numbers are randomized, that is, they follow no logical or coherent order, but with the restriction that a number can appear once and only once in any column or row.

There is a reason for this. We suspect that there will be a substantial amount of learning in this experiment, and we want to be sure that the various inclinations get a fair test. If we tested all the 0° inclinations first, we would probably get values which are too high because the

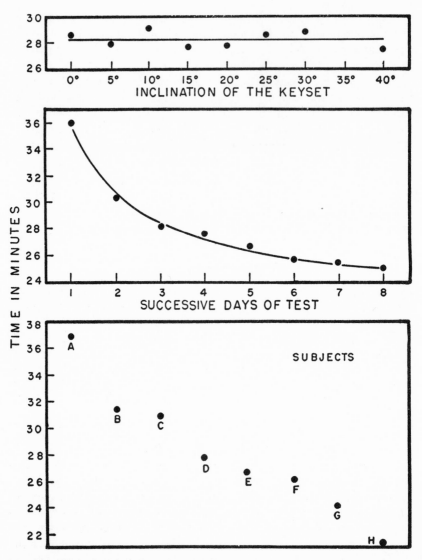

FIG. 29. The uppermost chart shows the average times required by eight operators to key 150 ten-digit-letter combinations with a keyset inclined at various angles. The middle chart shows the average times required by the eight operators to key these combinations on each of eight successive days. The lowest chart shows the average times required by each of the eight operators to key the same digit-letter combinations. All three sets of values come from a recombination of data obtained with the experimental design shown in Table 34. (Unpublished data from an experiment by Scales and Chapanis[97] from Chapanis.[27])

subjects had not yet become familiar with the apparatus. This would introduce a systematic bias in the data. The solution, you will notice, is to test each inclination first with one subject, second with some other subject, and so on. Thus if there is any systematic practice effect in this experiment, the effect should be distributed equally throughout all of the angles, and the net result should be to cancel out this factor.

There are some other details of this experiment which are important, but which are a little aside from the point of this discussion. However, we can mention that the lists of numbers to be keyed were all carefully prepared in advance and that they were different for each subject on every day. Carefully prepared instructions were read to the subject, and each subject was given two days of preliminary trials to become familiar with the apparatus and test procedures. The time recorded for each subject is the total time required to key 150, ten-digit-letter combinations.

Results. The results of this experiment are shown graphically in Figure 29. You will note that there are three sets of findings here. The top set has to do with the effect of inclination. To get these values we simply averaged the keying times for the columns in Table 34. The second set of data is concerned with the effect of practice. To get these entries we computed an average keying time for all those places where a 1 appears in Table 34, then an average keying time for all those places where a 2 appears in Table 34, and so on. The final set of data in Figure 29 shows individual differences. These data were obtained by averaging keying times across rows in Table 34.

The interpretation of these data is relatively straightforward. The largest source of variation in these data comes from the differences between subjects. Subject *H* is much faster than the others, and the averages for the other subjects scatter over a large range of values. As we suspected there might be, there is a pronounced effect due to practice. This is the second largest source of variation in the data. The total spread of daily averages is not quite as large as that found between subjects, but it still represents a fairly large amount of change. As compared with these two sources of variation, the angle of inclination of the keyset has a trivial effect, and there is no basis here for believing that a tilted keyset would affect operator keying time.

The statistical analysis of the data. The appropriate form of statistical analysis for the experimental design in Table 34 appears in Table 35. Perhaps the first thing you will notice about this analysis is that it has no interactions. In general, there can be no interactions unless trials have been scheduled for every possible combination of the main variables. For example, *Operator A* was tested on *Day 1* with the *keyset at 25°*. If we wanted to test all the other combinations of these three variables we would also have to test *Operator A* on *Day 1* with the *keyset at 0°*; *Operator A* on *Day 1* with the *keyset at 5°*; and so on. But there is only one Day 1, and there is a definite limit to the amount of testing you can schedule for any one sitting. Even if we could schedule several hours of testing in one day, the subject will have learned quite a bit during the first hour. Thus, in this situation it is impossible to test all combinations of the three main variables. This is generally true of experiments in which learning, or performance on successive days, is

TABLE 35. *Basic form of the analysis of variance for the experimental design shown in Table 34*

Source of variance	Degrees of freedom
Between subjects	7
Between inclinations of the keyset	7
Between days	7
Residual	42
Total	63

one of the main variables. A subject can start out fresh only once for any particular learning task.

We might also point out that the residual variance in this experiment is composed of different elements than the residual variance in every other example discussed so far. In this case, the residual is made up of pure random error plus the effects of the interactions which we cannot untangle and measure. Statisticians say that the interactions are confounded in this kind of design. What they mean is that the interactions are mixed up with other sources of variance and cannot be separated.

This kind of experimental design is called a Latin square and it serves a useful function in human engineering studies. By careful design of our experiment, we were able to get good data on a particular point (the effects of tilting the keyset) even though we knew, or suspected, that there would be other major sources of variation (learning, and differences between subjects) in the experiment. *Rather than try to hold these other factors constant, we deliberately designed them into the experiment in such a way that we could evaluate their relative importance.*

Another illustration of a Latin-square design. The experiment on the toll-operator's keyset was deliberately picked to illustrate the Latin-square design because (a) it is a nonmilitary example, and (b) the data are so handy. Comparable examples can be found in other applications. For example, Leyzorek did an experiment on radar which is an almost exact parallel of this one on keysets.[72] The keyset experiment made use of an 8 × 8 Latin square—so called because there are 8 rows and 8 columns (see Table 34). Leyzorek used a 7 × 7 Latin square since he tested only 7 radar operators and mounted the face of a radar at 7 different angles. Parenthetically, Leyzorek's findings also closely duplicate the findings on the tilted keysets: He found a significant effect due to learning, significant differences between radar operators, but no significant effect as a result of the angle at which the radar was tilted.

Some summary observations on the Latin-square design. Perhaps the most important characteristic of the Latin-square design is the elegant counterbalancing scheme it uses. This in turn makes for an extremely efficient design in the sense that it enables the investigator to test several values of three basic variables with an extremely small number of trials.

This efficiency is gained at some cost, however. First and foremost is the loss of all interactions, and in human engineering work these may be very important. Some statisticians in fact recommend that the Latin-square design not be used whenever there is reason to suspect that there may be significant interactions in the data. The second difficulty with the Latin-square design is that the statistics are a little murky. The fact that the error term is a heterogeneous mixture of all sorts of variances makes it difficult to give a clean-cut interpretation to the data. When the effects are huge, however, as they were in Figure 29, this does not constitute a serious problem. Despite these disadvantages, the Latin square is an important one for the human engineer to know, especially when he wants to get a quick test of several interesting variables.

One other important use of the Latin square is in counterbalancing auxiliary details in an experiment. An illustration of this usage was the way in which targets were assigned to the various test conditions in

Table 30. In that case, the Latin-square principle was used, not as a basic form of experimental design, but simply as an effective way of counterbalancing for possible differences between targets.

OTHER EXPERIMENTAL DESIGNS

In this chapter we have discussed four major classes of experimental design: (a) experimental designs for making simple comparisons; (b) factorial designs; (c) treatments $\times$ subjects designs; and (d) Latin-square designs. Together these make up a solid basic repertory of experimental designs which should enable you to handle most kinds of experiments in human engineering. They do not, however, exhaust the list of available designs, and for a more extended treatment of this subject matter you should consult more advanced books by Cochran and Cox,[36] Kempthorne,[68] and Lindquist.[73]

HOW MUCH REALISM DO YOU BUILD INTO EXPERIMENTS?

An important consideration, and one of the most difficult the experimenter may be called upon to resolve, is the question of how much realism he should build into his experiments. This problem is identified by various names—in some laboratories it is called the problem of simulation—although the name is not really important. The point is that it is a problem common to all of applied science everywhere. However, it is a problem which concerns engineers much more than laboratory scientists. The reason for this is not hard to see. The results of tests which the engineer makes may be used to decide about a production change in the manufacture of thousands, or even millions, of items. Or, the results of engineering tests might be used as a basis for design changes in an aircraft, tank, ship, or electronic device where, although the quantities may not be great, the expense is. In both of these, and in many related kinds of example, a wrong decision may be very costly. So it behooves the engineer to be sure that his tests give him correct answers—answers that will not result in disgruntled customers, decreases in sales, or in failures and accidents.

To see more clearly what the issue of realism is, let us consider an experiment designed to evaluate the readability of two different kinds of aircraft dials. How should you make the tests? Can you compare two cardboard dials? Should you use two real dials? Should you surround them with other dials? Is it necessary to put the whole assemblage into an aircraft cockpit? Is it necessary to conduct the tests while the aircraft is actually in flight?

Each of the above questions is a question about successively greater amounts of realism and there are, of course, many intermediate degrees of realism which we have not identified. In addition, realism extends to factors other than that of the equipment itself. It applies, for example, to the kinds of subjects you use for your tests. The tests might be done on high-school girls, high-school boys, college boys, military enlisted men, or military pilots. Each of these samples represents successively greater realism to an operational population.

THE ARGUMENTS FOR HIGHLY REALISTIC EXPERIMENTS

What are the pros and cons on this question of realism in experiments? The most important argument for maximum realism is that we have the greatest confidence in the results of such experiments— that is, we have confidence in them if they are, or can be, done well. The greater the realism of our experiments, the more certain we are that they will tell us exactly what will happen in real-life situations.

You cannot control all the variables in an experiment. What is the basis of our confidence in realistic experiments? This point is sufficiently important, and sufficiently basic, that we might dwell on it for a moment. When we talked about the method of experiment we spent a considerable amount of time talking about control—the control of relevant and irrelevant variables. This principle is easy enough to state, but as a practical matter it is impossible to put into practice. There are two things wrong with it. The first is that if you tried to control all the factors you were not interested in, you would never be able to do a human experiment. You have to worry about temperature, humidity, the time of day, the day of the year, the past history of your subjects, and so on. The job of trying to control these and other factors is so immense that you would never be able to start a human experiment.

The second practical difficulty is that we are never really quite sure what is relevant or irrelevant, i.e., what are all the important variables to control. In human experimentation we have not yet identified them all. Indeed, our ideas about what needs to be controlled are progressive—evolving with the state of our knowledge. Factors which were not controlled in experiments done even twenty years ago, we know today are quite important. That is one reason why scientists have to recheck constantly the results of old experiments.

These practical difficulties are so great that we cannot help having some residual doubts in applying laboratory data to real-life situations: Did we really think of everything important in the laboratory experiment? Did we consider all the relevant variables? Ultimately, our only validation comes from field trials, or real-life tests.

Experiments are artificial situations. Aside from the difficulties of controlling variables, there is another, quite different reason for making human engineering experiments as realistic as possible. The typical laboratory experiment is a highly artificial situation, which the subjects perceive as such. As a result, subjects do not behave the same in the laboratory as they would in a real-life situation. The behavior of a gunner aboard a destroyer is quite different from the behavior of the same man when he is operating a gunnery simulator in a laboratory. The behavior of a pilot flying an aircraft is different from his behavior when he is "flying" a jet trainer on the ground. Since we usually have no way of knowing how much this artificiality changes the behavior of our subjects, we are on safer grounds if we make our experiments as realistic as possible. In this way we can feel more certain about the applicability of our findings.

Another aspect of the artificiality of laboratory experiments is the fact that when you take a problem into the laboratory, the variables may not be the same as those in real life. In trying to get control over variables in the laboratory, the experimenter may have to impose such restrictions and rigid conditions that the variables are distorted from what would be expected under natural conditions. At best, a laboratory experiment is only an approximation to life. For human engineering purposes, the closer we make this approximation, the safer we will usually be.

THE ARGUMENTS AGAINST REALISM

The most important argument against maximum realism is expense. Conversely, the most important argument for "artificial" laboratory experimentation is that you can get good answers at relatively trivial costs. To run adequate tests on two different aircraft dials in actual flight might easily cost a half-million dollars. If this figure seems extravagant, you must appreciate that it includes, among other things, the cost of constructing the two dials, the cost of installing the dials in, let us say, two aircraft (which calls for skilled technicians and elaborate procedures), the cost of flying the aircraft for as many hours as are needed to obtain sufficient data (it costs as much as $10,000 merely to get some modern aircraft airborne a single time!), and the salaries of the pilots who will fly the aircraft and perhaps the skilled observers on whom data will be collected. Compared with this figure, a laboratory experiment on the same problem yielding data of equal reliability, could realistically be done for the cost of getting a single plane off the ground once. The cost of any kind of research is always much greater than most people think, but the difference between the costs of laboratory and real-life experiments is usually enormous.

The difficulty of controlling variables. A second, and very important, argument against real-life experiments is a technical one. Basically it is this: In realistic experiments, the experimental situation usually contains many uncontrolled sources of variability which contaminate the data in which the experimenter is primarily interested. As a result, to get comparable levels of statistical significance in results, realistic experiments usually require many more trials than laboratory experiments (refer also to the discussion of this point on page 163). Let us see if we can illustrate the basic idea involved here. In a laboratory experiment on dial-reading, it is no trick at all to control the illumination, and the experimenter would undoubtedly do this, because the level of illumination affects markedly the readability of dials. If the experimenter wants, he can easily run his tests on the two dials at several levels of illumination using a multi-factor design of the type we talked about earlier. But whatever he does, he is sure that the conditions under which dial A is read are the same as those under which dial B is read. He has *controlled* the illumination and has insured direct comparability of his results.

But now let us consider the same experiment in aircraft. Suppose that tests are actually done in flight. In all probability this means that the tests are thereby done under varying conditions of cloud cover, sky brightness, shadowing, reflections from the interior, direction of the aircraft with respect to the direct rays of the sun, time of day, and so on. Now see what happens. During the flight when the pilot is using dial A, the sky brightness might be that of a clear cloudless sky at mid-morning. During the flight when dial B is being used, there might be a slight haze. Or under one set of tests, the plane might be heading away from the sun; under another set of tests, the heading of the plane might be a little off the previous course so that faint shadows appear on the dial. The illumination, in short, cannot be controlled throughout all tests. It cannot even be controlled from minute to minute. As a result, the only way the experimenter can be sure that the results for dial A are really comparable to those for dial B is to run such a large number of tests that he can be reasonably confident that the average condition of illumination for one dial is matched in the tests with the other dial. This is what we mean by an uncontrolled source of variance influencing the test results.

One solution which might appear to get us out of this predicament is to mount the two dials side by side so that they receive the same amount of illumination at any moment. Then if the two dials are read nearly simultaneously in time, we have some control over the illumination. But is this a solution? Would we have a fair test of the readability of dial B if the pilot read it immediately after having read dial A? No, because the dials would have the same reading and now we run into the problem of memory influencing what the pilot says. If we separate the observations in time to get around this problem we run right back into the possibility that the illumination will change between observations.

Now recall that this is only one factor. In realistic experiments there are many more such uncontrolled factors contributing to contaminate the test results—variations in temperature, oxygen deprivation, and other environmental factors which may alter a man's performance from time to time; variations in the emotional condition of the observer (after all, taking a plane into the air *is* a stressful task which varies from flight to flight); variations in the performance of the vehicles; variations in the traffic density from moment to moment; and so on. All of these uncontrolled sources of variation in test conditions mean, in effect, that highly realistic experiments are generally inefficient for teasing out the particular relationships in which you may be interested. Furthermore, having many uncontrolled sources of variation in our tests may always leave us wondering whether we really ran enough trials to balance out all these factors for the comparison under test. The net result is that our confidence in the results of realistic experiments may sometimes be an illusory one.

The statistical consequences of uncontrolled variables. The preceding discussion about the effects of uncontrolled variables can be made more concrete by tying them into their statistical consequences. When we were discussing the statistical analysis of multi-variable experiments in Chapter 4 we showed that there was always some residual variability which we called random error. This random error was used to test the significance of the effects we were interested in. To do this we, in effect, compute a ratio between the variance attributable to any main effect (say that between methods in the Gebhard experiment) and the random-error variance. The higher the ratio the more likely we have a significant effect; the lower the ratio the more likely we have a random effect.

But now let us see what happens to these variances if we add uncontrolled sources of variability to our experiment. These additional sources of variability will not increase the difference due to the methods as long as they are distributed randomly with respect to methods. They will, however, increase the denominator of the fraction, i.e., the variance which we have computed and identified as random error. Thus, the effect of having uncontrolled sources of variability in an experiment is to increase the error variance. This, in turn, decreases the efficiency of the experiment, because it makes it harder to establish the significance of an effect which may be produced by one of the main variables.

The stability of comparisons. Another consideration is that, for many kinds of tests, increasing the realism of the experimental situation does not change the relative values obtained in the laboratory. Another way of saying the same thing is that it is relatively easy to find out if *A* is better than *B*, and the stability of this finding is generally pretty good if it is based on a sound laboratory experiment. It is much harder to find out "how much better *A* is than *B*," and the stability of this kind of answer requires very careful field testing.

To look at a real example, Grether was interested in comparing the relative effectiveness of several kinds of aircraft altimeters.[61] The results he obtained on six are shown in Figure 30. Notice that he used two kinds of subjects—Air Force pilots and college students. In terms of the realism issue, it is more realistic to use pilots than students. Note that there are differences between the results obtained with the two groups of subjects. But here is the important point: The dial which was best for the students was also best for the pilots. The dial which was poorest for the students was worst for the pilots. The relative rankings of the dials are, for all practical purposes, identical for the two classes of subjects (the actual correlation between the two sets of ranks is +0.9), although the actual absolute values varied. Discrepancies between the two occur only for those dials where the results are so close together that they may be considered equivalent.

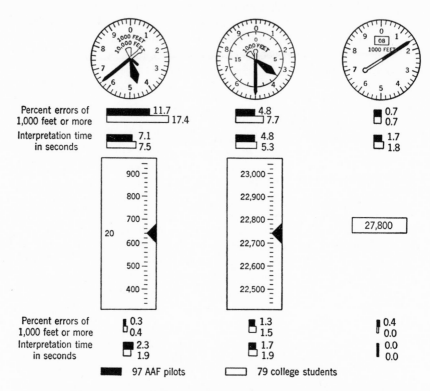

FIG. 30. The conventional three-pointer altimeter (*upper left*) and the five best experimental altimeters found in a study by Grether. The number of large errors and the average reading time for pilots and college students are shown beneath each dial. (Data of Grether[61] from Chapanis, Garner, and Morgan.[29])

To look at another illustration, Chapanis did a series of three studies on the relative efficiency of a bearing dial and bearing counter for reading the bearings of targets from radar scopes.[21, 22, 23] These three experiments had increasing amounts of realism. The first was a highly artificial experiment with a mock-up of a bearing counter and with the subjects merely reading numbers from a dial or counter. The second was a little more realistic, because operators were required to read the bearings of simulated targets with a bearing counter which actually worked. The third experiment was still more complex, involving a number of other variables with a group of men using many radars. The results show that the bearing counter was better than the bearing dial in every experiment. Increasing the realism did not change the relative rankings of the two instruments.

Along the same lines, Weldon and Peterson found that a dial booklet test gave the same relative rankings for five dials as did experiments done with subjects reading genuine dials.[112] As far as readability is concerned, the two methods gave equivalent results. They point out, however, that the paper-and-pencil test provided no information about mechanical problems of manipulating the controls and associated dials. Similarly, Churchill found that mean reading times and errors were almost exactly identical for reading panels of genuine dials and for reading dials which were projected onto a screen by means of a lantern slide projector.[34] Finally, Bessey and Machen[12] found that values of room illumination and radar screen brightness which had turned out best in laboratory studies[114, 115] were also best when they were tested under operational conditions at an RCAF radar station.

To sum up, then, there is a substantial amount of experimental evidence to show that the results of laboratory experiments can be safely applied to more realistic situations. We are sure that this is not universally true, and, unfortunately, we are not sure just exactly how far we can trust this general rule. In so far as it holds, however, it does give us increased confidence in the value of laboratory studies.

The impossibility of controlling motivational factors in real experiments. The final point we should note is that experience shows it is extremely difficult to do good experiments with real ships, airplanes, tanks, and related complex systems. It has been tried on numerous occasions, but the results are generally disappointing because of the extreme difficulty of controlling variables the way they need to be controlled. The reasons for these failures are seldom discussed, partly because experimental failures do not get written up and reported. In addition, massive experiments involving groups of ships, aircraft, or other military components are usually attempted to evaluate equipment which is classified. Finally, the experimenter may frequently not know that the data are seriously biased. It may be revealed only to someone who fraternizes with the "subjects" of the experiment in a bar.

One of the most difficult things to control in these real-life experiments is the complex interplay of the motivations involved. Perhaps the simplest way of illustrating the problem is to discuss some genuine, but slightly disguised cases. Here, for example, is an "experiment" to test two alternative bombing systems. However much the experimenter tried to get honest results, the crews of the aircraft knew, or strongly suspected, that the results of the "experiment" would be used to evaluate their efficiency. Although no one likes to admit it, the fact of the matter is that "bribes" and other rewards exchanged hands between aircrews and the people collecting the data. The crew of the aircraft was not interested in any experiment; its primary interest was in avoiding a bad bombing score.

Here is another experiment designed to evaluate the comparative efficiency of two methods of using sonar in anti-submarine work. By devious means the executive officer of the surface ship managed to get the test schedule showing where the underwater targets were supposed to be at various times. His primary concern was in how his ship showed up.

Still another experiment was undertaken in a factory manufacturing consumer goods. The purpose was to evaluate the comparative effectiveness of two machines. The plant was unionized and the test subjects deliberately slowed down to make the newer method show up poorer than it really was.

The last example comes from a massive, real-life experiment involving many observers who collected data without supervision. Since observations sometimes had to be made under unpleasant and inconvenient circumstances, some of the observers discovered that their tasks were much simpler if they manufactured the data at home between television programs. It was only by chance that the principal investigator discovered this deception.

Controlling the motivation of test subjects in a laboratory is difficult enough. It is infinitely more difficult when these experiments are done under conditions where the outcomes of the experiments may affect the status, rank, or earnings of the subjects involved. The complex interplay of motivations in real-life situations is amply illustrated in the case studies cited by Roy[96] and Dunlap[43] from actual industrial experience. Although these case studies do not describe experiments, several of them show clearly how the advantages of technological improvements and human engineering changes may be more than offset by deliberate "slowdowns" and even outright deception. In every instance the total context of the situation produced results which had not been anticipated by the engineers introducing the changes. In fact, it is questionable whether any one might have predicted the results that actually occurred.

The final point to be made is that this problem concerns more than just the workers, or subjects, in the experiment. It may be equally difficult to control the motivations of the other participants—the foremen, supervisors, or observers—and to get the kind of co-operation which is essential to good experimentation.

REALISM IN EXPERIMENTS: SUMMARY

To sum up, there is no final answer to the question of how much realism one needs in experiments. In the final analysis, the decision rests upon the good judgment of the experimenter and upon his knowledge of what is, or is not, likely to be relevant to the particular relationship he is looking for. Realistic experiments give us greater confidence in the applicability of our results. On the other hand, realistic experiments are hard to do well, they are enormously expensive, and they are highly inefficient because they may actually introduce so many uncontrolled sources of variance that important relationships may be effectively concealed. Finally, highly realistic experiments are often unnecessary because they fail to change the basic result anyway. Perhaps the best strategy is this: Start out with laboratory experiments *using as much realism as you think is justified in the light of your best judgment about the important relevant variables.* Whenever laboratory experiments give you results which are positive and likely to be put into practice, then is the time to give the laboratory result a field trial to determine the stability of the laboratory findings in a different context.

Chapter 6

Some special problems in

experimenting with people

By now it should be apparent that there are a number of ways in which man's complexity can bias or distort the outcome of experiments you do with him. The preceding chapter was primarily concerned with methods of designing experiments to guard against such biases. In this chapter we shall continue along the same lines and, in particular, shall consider some special problems of human experimentation in roughly the same order you would meet them if you were designing an experiment. You will not always have to face all of these problems, because, of course, experiments differ enormously in their complexity. Some are relatively straightforward; others exceedingly complex. In addition, for different problems some of these questions are more important than others. About all I can do here is to identify some of the persistent problems and let you decide which you should consider in planning any particular experiment.

WHAT IS THE QUESTION?

The most difficult part of most experiments is deciding precisely what the problem is. What is it you want to find out? In general this means identifying three separate classes of variables: (a) the independent variables; (b) the dependent variables; and (c) the variables to be held constant.

ON INDEPENDENT VARIABLES

The independent variable is the thing that is deliberately varied because you want to find out what happens when you change it. Here are some examples of independent variables:

1. Variations in dial design.
2. Different inclinations of a control console.
3. Shape-coded versus uncoded control knobs.
4. Successive days of training.
5. Variations in ambient temperature.
6. A steering wheel control versus a joystick.
7. Frequency-selective versus wide-band circuits for auditory communications.
8. Different amounts of illumination.

There may be more than one independent variable in any experiment, of course. For example, we may be interested in testing two different dials, but at the same time be interested in finding out how large the differences are between dials when we test different sizes of these dials. Thus, we have here two independent variables: type of dial and size of dial. We could add a third variable by asking whether there is any difference in the performance of subjects on these dials as a result of variations in illumination. In the experiment on tilted keysets cited earlier (page 193) there were three independent variables.

Uni-dimensional variables. As you can see from the list above, variables may be continuous along some single quantitative dimension, or discrete, that is, qualitatively different. The inclinations of a control console are selected from a continuous variable—the console could theoretically be tilted to an infinite number of different inclinations all differing in this one respect. Time, in Figure 27, is also a continuous, uni-dimensional variable. When the independent variable is quantitative along such a dimension, we can plot its values along the baseline; plot values of some dependent variable along the ordinate; and get a functional relationship between the two. Figures 27 and 29 illustrate such functional relationships.

Ordinarily the most important problem in selecting points for test from a continuous dimension is that of selecting enough points so that you can plot the relationship with some confidence. At the very minimum, three values should be selected for test if you have good reason to suppose that the functional relationship will be linear or some simple monotonic curve (as in Figure 29). On the other hand, if you have reason to suppose that the functional relationship will be complex, as illustrated in Figure 27, then five points are about the minimum number which will suffice. These figures of three and five are absolute minimum values. Ordinarily you will need more than these to plot functions with confidence.

The *range* of test points is the second problem which is important in selecting values from a continuous dimension. What the experimenter most generally wants to do is to use a range which is wide enough to cover all of the potentially interesting region of the independent variable. Sometimes previous research will suggest what this range of values should be. In dark-adaptation studies, for example, we know that a range of 0 to 30 or 40 minutes is ample for most experiments. Dark-adaptation changes so slowly, and so little, after 30 minutes that durations longer than this are of little interest.

Sometimes a range of test values can be selected on an *a priori* basis. If one of your independent variables is the amount of illumination falling on dials to be used in a submarine control room, you can pretty well decide in advance that a range of illuminations from 1 to 100 foot-candles will be ample. Seldom do you find indoor panel illuminations outside of this range.

Sometimes, however, the only way to decide on a suitable range of test values is to do a small pretest from which you can get some insight into the region most likely to be of interest. If you were starting out to do a study in a new area of research—say, for instance, the effect of various amounts of transmission lag on the accuracy of pursuit tracking—you would probably find it worthwhile to try lags of say 1, 5, 10, 20, 40, and 80 seconds. A few trial runs would undoubtedly show that a human operator cannot even begin to track most targets if the time lags exceed T seconds. This quickly narrows the range of potentially informative values to those under T seconds.

Having selected a *range* of test values the last question to be decided is their spacing. Here one important rule to note is that in most quantitative sensory dimensions the human being responds to *ratios* of intensities rather than to arithmetic increments. This means that if the independent variable is amount of light, loudness, or frequency of sound waves you will probably find it best to space (and later to plot) the test values on a logarithmic scale rather than a simple linear one. To illustrate with the submarine illumination problem mentioned above, suppose that we want to test five intensities of illumination between 1 and 100 foot-candles. A reasonable set of values would be 1, 3.2, 10, 32, and 100 foot candles.

Qualitative variables. Some variables in human engineering studies are qualitative—that is, they cannot be plotted along any single dimension or scale. For example, the altimeters in Figure 30 are qualitatively different. Qualitative variables often give us trouble if we are not alert to some special problems in their selection. Perhaps the best way of illustrating these is by an example.

When you tune your radio you turn a knob which, on some radios, moves a pointer along a linear scale calibrated for frequency. Let us suppose that we are concerned with the question: Is a crank or knob better for making such settings? We construct a linear scale and pointer and then fix up a knob and crank so that they can be used interchangeably with the display. We pick a knob and crank of the same diameter and with the same gear ratio between the control and the pointer—one complete revolution of the knob or crank moves the pointer one-half inch. Let us say that we measure the time it takes subjects to position the pointer (within certain limits of accuracy) to different values on the scale. Having arranged these and other conditions of the experiment, we run the tests and find that it takes less time for our subjects to make settings with the crank than with the knob. Statistical tests show the difference to be reliable.

Let's look at this finding carefully. Have we shown that a crank is better than a knob, or have we merely shown that this particular crank is better than this particular knob? To see more clearly the issue involved look at Figure 31. This chart shows data of exactly the sort we have been speculating about. Notice that for a control ratio of 0.5, the crank is in fact faster than the knob. But if we had happened to pick a control ratio of 1.0 for the two controls we would have found that they were about equal. Finally, if we had happened to pick a control ratio of 5.0 we would have found that the knob is better than the crank. What value should we have picked? If we are limited to one value for each control mechanism and want to get an answer of the greatest possible generality we should have picked the *best* control ratio for the crank, and the *best* control ratio for the knob. If we had done this we would have arrived at the conclusion that the knob is faster than the crank when each is operated at its best control ratio. What we have said about control ratios applies as well to the sizes of the controls, their shapes, placement, and all the other variables which affect their operation.

All this suggests a principle of some importance: When you compare qualitatively different instruments, devices, or methods, it is usually essential that you compare the best possible samples of each one. This means that you should select the best values of all those variables which are to be held constant in the experiment.

This principle seems obvious enough as we have discussed it for this simple application. But it is not always so obvious when it crops up in other kinds of human engineering experiments. For example, an experiment was devised to test several different kinds of tank-steering con-

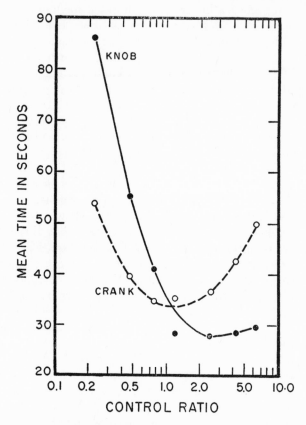

FIG. 31. Average times required by three subjects to make pointer settings on a linear scale with a knob and crank. The abscissa shows the various control ratios— inches of pointer movement per revolution of the knob or crank. (Data of Jenkins and Connor[66] from Chapanis.[27])

trols: a wheel, a joystick (similar to that used in airplanes), and a two-lever braking system (the tank turns to the right when the operator pulls on the right-hand lever and stops that set of tank treads; it turns to the left when the operator pulls the other lever). Since this was an expensive experiment to set up, the plan called for testing only one sample of each kind of control. The experimenters were interested in a very general answer: Is control A, B, or C the best? In this case, it was essential that the best possible control of each type should be tested. This means that the best size, torque, friction, inertia, and placement of each control had to be selected for test. Only in this way could the experimenters be sure of finding out what they really wanted to know.

To use another example, how should we evaluate the results of studies done to compare different kinds of radars? Many studies of this type have been done on prototype models—on the only radar of each type available. Since we have no assurance that the investigators tried to maximize the performance of each radar before the test began, we are left with the nagging doubt: Would the results have come out differently if the experimenter had modified the radars in this way or that?

This principle becomes especially difficult to handle in the comparative evaluation of complex man-machine systems in which many men may be involved. Suppose that we want to test two alternative methods of organizing an air traffic control center by measuring the effectiveness with which each type of organization can receive, process, and plot information about aircraft within a radius of 200 miles. We can have confidence in the generality of our findings only if we can be sure, among other things, that our teams of men were working at peak efficiency. If this requirement is not met, we may be unable to evaluate certain outcomes of the study. For example, suppose our tests show that system A can handle information better than system B. But now let us suppose that the men who worked with system A were a smooth-functioning team, whereas the men who worked with system B didn't get along together for some reason or other. Under these circumstances we cannot be sure whether the superior performance of system A is due to the fact that (a) it is really better, or (b) the men who operated it were more efficient than the men in the other group.

One alternative to picking the best possible control (system or device) is to test each control throughout a range of values to locate optimum points, as was done in Figure 31. Functional relationships of this type give us answers of greater generality than can be obtained by picking only one set of parameters for test.

On the treachery of certain words. A particularly troublesome set of variables are those which deal with certain human characteristics.

Suppose, for example, that you wanted to do an experiment in which fatigue was one of the major independent variables. Operator fatigue is a factor of considerable importance in many man-machine systems and as such is a proper subject for experimental inquiry. But how do you vary fatigue as an independent variable?

Everyone knows—or thinks he knows—what fatigue is because he has experienced it so often. Yet the man who is exhausted after a hard day at the factory or the office may go home and play a round of golf or some sandlot baseball. The facts are that fatigue is one of the most difficult human characteristics to measure because it is so elusive. You seem to have fatigue in one particular situation, change the situation a little, and the fatigue has vanished. Part of the problem arises from the fact that fatigue is apparently a number of different things, but we have only a single word to designate them all. As a result it is easy to fall into the trap of supposing that everything which goes by the name *fatigue* is the same. The "fatigue of radar operators who watch blank radar scopes for hours at a stretch" is quite a different kind of phenomenon from the "fatigue of infantrymen who have to carry a 30-pound pack around some hills with machine guns spitting at them."

Other troublesome words of this general type are *stress, alertness, carelessness,* and *accident proneness*—all legitimate subjects for scientific study, but difficult to quantify, measure, and control.

In dealing with such concepts the first precaution to observe is to define your concepts operationally (refer again to the discussion of this matter on pages 19 and 20). Don't try to do an experiment on "learning"; do an experiment on "learning to strip a Browning automatic rifle," or "learning to fly a T-33." Each of these learning tasks differs from the other, and they both differ from "learning integral calculus."

The second point to observe is that once you have given an operational definition of the terms you aim to investigate, there is the very real question, "Is it as general as you think?" Having a name for some human capacity, it is easy to fall into the error of believing that the name means more than it really does. We tend to categorize and generalize human capacities and traits when, unfortunately, most of them are really much more specific than we believe. We like to believe that people are honest or dishonest; that things are black or white. Actually, even the most pious churchgoer is dishonest about some things, and the blackest scoundrel has his own sense of honor—distorted though it may seem to most people. Things, unfortunately, are not absolutely black or white; they are usually some shade of gray in between. With our present state of knowledge about human behavior, you will be on much safer ground if you automatically go on the assumption that most kinds of human behavior are highly specific. Psychologists have some *general* laws of human behavior, to be sure, but if the psychologist you talk to is a good one, you are almost certain to hear him say, "Yes, in general that's true, but in this situation. . . ." In short, don't extrapolate too far beyond your operational definition.

ON DEPENDENT VARIABLES

In contrast to the independent variable, the dependent variable is usually much more difficult to define and agree on. The dependent variable is what you intend to measure—the criterion. Two useful criteria in human engineering research are time and errors—the time it takes to do something, and the number of errors or amount of the error made in doing it. Both are objective, easily measurable, and defensible in many practical situations.

One trouble with error data, however, is that in some situations the error frequencies may be very low. Thousand-foot errors in reading an aircraft altimeter occur seldom, yet each represents an extremely important and serious kind of error. When error frequencies are low, experimenters frequently arrange test conditions which are more difficult than normal so as to deliberately increase error frequencies. The assumption here is that if variable *A* (or situation *A*) is better than *B* under stressed conditions, it will also be better under normal conditions. Ways of stressing the subject, and thus of increasing the difficulty of the task are by:

1. Speeding up the presentation of test conditions.
2. Presenting the test under adverse environmental conditions, e.g., under poor lighting.
3. Presenting distracting, annoying, or irrelevant tasks which the subject must also perform.

If you use one of these methods you should clearly recognize the assumption underlying it. Although it may be tenable for most experimental situations, it is not necessarily valid for all.

Other dependent variables. Three other measurable and useful dependent variables for human engineering studies are: (a) the amount of work done per unit time; (b) the amount of time or number of trials required by novices to learn how to use a piece of equipment; and (c) psychophysical criteria. The first of these needs scarcely any comment, but we might say a few words about the other two.

It is characteristic of equipment which has been properly designed from the standpoint of its operator that people can learn to use the equipment much more rapidly than equipment which has not been properly designed. Thus, learning time is a useful and practical criterion, or dependent variable, for evaluating alternative devices or methods.

The next chapter will show how psychophysical criteria can be used as a dependent variable in human engineering experiments. The experiment on fuzziness in TV pictures as a function of bandwidth (page 264 ff.) is a case in point. Experiments on radar detectability commonly use thresholds as dependent variables.[8, 10, 113, 114] In the experiment by Egan *et al.* a "threshold of perceptibility" was used as the dependent variable in evaluating the efficiency of frequency-selective circuits in a multi-channel communication problem.[46]

Problems in the selection of a criterion. The dependent variables listed above are some which have been useful in a variety of studies. However, they are only a few of many possible ones. Selecting the appropriate dependent variable, or criterion, is not always as easy as you might think. For one thing, a criterion which is relevant for one situation might not be for another. Suppose, for example, that you wanted to compare the merits of several voice communication systems. If these systems were designed for use in military tanks it might be entirely appropriate to evaluate the systems in a noisy environment by means of an articulation test.* On the other hand if you are working for a telephone company you might very well want to pick another criterion instead of, or in addition to, articulation test scores. Such a criterion might be the "naturalness" of the sound of a person's voice. When you call up your wife or mother after a long absence, perhaps the most important thing to her is that you should *sound like* yourself. This is a reasonable and justifiable requirement to put on a system (see, for example, David[39]).

An important point about this business of criteria is that different criteria may lead to quite different test outcomes. Take the matter of deciding among several voice communication systems. Accuracy of

* Articulation tests measure the proportion of words, or sentences, that can be heard correctly when they are transmitted over a communication system (see Chapter 8).

speech reception in a noisy environment is probably one of the most important requirements of a system if it is to be used in tanks. One way of increasing the intelligibility of speech in noise is to subject it to amplitude distortion of a type known as peak clipping. This, as its name suggests, distorts the speech so that it no longer sounds natural. But under certain conditions (see Chapanis, Garner, and Morgan,[29] pp. 232 ff. for a discussion of this point) it may make the speech much more intelligible in noisy places. So if the criterion in your tests is the accuracy with which speech is transmitted, a system with peak clipping might turn out best. On the other hand, if your criterion is the *naturalness* of the sound of transmitted voices, a system with peak clipping might turn out worst.

What you really want to do, of course, is to select a dependent variable which is relevant to the application you have in mind. If you are trying to design a dial so that it can be located easily on a crowded display panel, this is what you should be measuring. If you want to design an airspeed indicator so that a pilot can maintain his airspeed within two per cent of a stated value, that is what you should be measuring. In any case, the choice of a dependent variable assumes that you have first answered for yourself the question, "Just what is it that I should be measuring?" A discussion of these and other problems in the selection of dependent variables for studies of instrument displays is contained in the report by Senders and Cohen.[99] Many of the things they say, however, can be extended to other areas of research.

Are you overlooking other important criterion measures? In the discussion above on the criterion of naturalness of the sound of a person's voice, I suggested that this might be used *in addition to* the results of articulation tests. This point needs some elaboration. Frequently there may be more than one relevant criterion for evaluating a man-machine combination. In such instances the experimenter may not be getting the whole story if he selects only one of them for test.

Two very useful kinds of mechanical indicators are the moving-pointer instrument and the direct-reading counter: the two indicators in the upper left-hand corner of Figure 30 are moving-pointer indicators; the one in the upper right-hand corner of the same figure is a combination of a moving pointer and counter; and the one in the lower right-hand corner is a simple counter-type indicator. A number of experiments,[21, 22, 23, 61] have shown conclusively that a counter-type indicator is superior to a moving-pointer instrument in terms of reading speed and accuracy. (Figure 30, for example, shows some data on this point.) But does this mean that we should change all moving-pointer instruments to counters? Not at all, because there are other relevant criteria which should influence our decision.

A counter is, in fact, better than a moving-pointer indicator *for reading exact numerical information from an indicator*. But mechanical indicators are used for three other principal purposes: for check reading; for setting information into a machine; and for tracking. Using all of these criteria to evaluate these indicators gives us a more complete picture of the comparative advantages and disadvantages of the two. Moving-pointer indicators are better than counters for check-reading and for tracking purposes; they are sometimes better than counters for setting information into machines and sometimes not, depending on the application. To summarize, in planning a human-engineering experiment it is well to look closely at the criteria which will be used as dependent variables to be sure that you have not overlooked some which might be critical in applying your findings.

Preferences as a criterion. Many human engineering studies use preferences as a criterion in evaluating alternative equipment designs. It is a very easy thing to ask subjects at the end of an experiment, "Which control (or dial, or panel) did you like best?" This is an interesting question to ask, and there is certainly no harm in asking it. In fact, it may occasionally lead to some useful information, but on the whole it is not nearly as useful as many people believe. What are some of the reasons why preferences are not a good criterion of equipment design?

1. Preferences are attitudes which are largely molded by the fashions, habits, and customs of the times and, as such, are highly unstable indicators of operational merit. That this is so is especially evident from the history of the design of many consumer items. For a long time, automobiles were black. People preferred them black. In fact, it was hardly conceivable that anyone would want to design an automobile in any other color. Now, to put it mildly, the fashion and times have changed.

Another illustration, more directly related to human engineering, comes from early experiences with electric lighting. When electric lights first appeared, most people actually *preferred* kerosene or gas lights. Now, of course, a large sample of preferences would give you exactly opposite results.

2. Preferences are often established on the basis of irrelevant factors. Being complex attitudes, preferences often arise from factors

other than the one you are interested in. In spite of what people say, their preferences and selections in automobiles are seldom based on engineering considerations—economy, efficiency, reliability, etc. Preferences for automobiles are often based on nonfunctional considerations—which have nothing to do with engineering merit—color, what the people next door own, what the boss owns, etc.

3. People tend to prefer the usual, the customary, the familiar. Often this shows up in strange and unexpected ways. Loucks, for example, made an extensive comparison of several types of aircraft attitude indicators.[74] As a good experimenter should, he ran his tests in counterbalanced order—some subjects were tested on *A* first and *B* second; others were tested on *B* first and *A* second. One of the interesting findings of the study was that subjects almost universally preferred the instrument they had been tested with last.

As another illustration, Sinaiko and Buckley describe an experiment performed at the Naval Research Laboratory on three different tactical display systems for use in naval combat information centers.[102] In one part of the study experienced naval officers from the fleet were asked to rate the systems in terms of their ability to handle targets. System *A*, using a conventional tactical display, was ranked best; System *K*, with a display similar to the conventional method, was ranked second; and System *M*, which had an entirely unique display, was ranked lowest. Apparently, familiarity strongly influenced the preferences of the naval officers, because objective data on target detection delays and target handling capacity showed that System *M* outperformed the other systems. Experiences like these show how easily one may be misled by preferences and how untrustworthy they may be.

4. The paragraph above showed that preferences are often uncorrelated with objective measures of performance. Subjects may also show strong preferences for a particular kind of design when careful objective tests show no differences between alternative designs. For example, in a study of toll-operators' keysets, Scales and Chapanis measured keying time and errors for keysets which were inclined at various angles with respect to the table top.[97] Although most subjects preferred a tilted keyset, keying time and errors were almost identical for all angles of inclination of the keyset. Exactly analogous findings came out of a study by Leyzorek on tilted radar indicators.[72]

5. Preferences may be established by the particular set of experimental conditions used for test and may be misleading for this reason. In one experiment, Tinker had a group of 144 subjects sit and read

under an illumination of 8 foot-candles for 15 minutes.[105] At the conclusion of the reading period the subjects were allowed to select from a wide range of illuminations the level which they preferred for reading purposes. The results are clear-cut and definite. The subjects preferred an illumination of about 8 foot-candles. But in the other part of the experiment, subjects read under an illumination of 52 foot-candles for the same length of time. At the end of this reading period they were again allowed to select the illumination they most preferred. Again the results came out with an apparently definite and clear-cut finding, except that this time they preferred an illumination of 52 foot-candles!

6. What people say they "prefer" or "like" may be what they think the "correct" or "expected" answer is. A number of years ago the Western Electric Company set up an elaborate study of factors thought to be important in productivity.[95] As one part of that study they attempted to find out the effect of variations in illumination on production. In one test two girls who wound coils were isolated in a special room and their productivity was charted throughout the tests.

First, the amount of light was increased regularly from day to day, and the girls were asked how they liked the change. As the light was increased, the girls told the investigator that they liked the brighter lights. Then for a couple of days the girls saw the electrician come in and change the bulbs, although in actual fact he took out bulbs of a certain size and replaced them with others of the same size, thus keeping the amount of light constant. The girls, supposing that the light had been stepped up further, commented favorably upon the "additional increase" in light. After a while, the experimenter started decreasing the amount of light from day to day, again asking for their reactions. After a period of such decreases, the bulbs were again changed without changing the amount of light. The girls now commented that the "reduced" amount of light was not so pleasant to work under and that they preferred more light. It is interesting to note, too, that production, the amount of work done, did not materially change at any stage of this experiment, thus providing another illustration of the point made in paragraph 4 above.

ON VARIABLES TO BE CONTROLLED

When we say that an experiment is well controlled, we mean that the experimenter has examined all of the possible relevant variables in his experiment and has tried to hold all of them (except the ones he deliberately designs into the experiment) constant. What are some of the ways of holding variables constant? Removing stimuli is one way. For example, using a soundproof, or lightproof, room is one way of insuring that extraneous sounds and sights (which vary from time to time) do not interfere with the observations you are interested in. Masking noises may also be used to control sounds that cannot otherwise be controlled. For example, if the apparatus makes certain noises which might influence the subject, the experimenter might arrange a masking noise to cover the offending sound. Subjects are often tested at the same time each day to control differences in performance between morning and afternoon. And so on. These rather mechanical factors are generally easy to control in an experiment. Some more difficult control problems are discussed below.

Motivation and its control. Motivation is perhaps one of the hardest variables to control in human experiments. In general, psychologists try to get subjects motivated by four different means:

1. Impressing the subjects with the importance of the experiment. Frequently it is possible to generate enthusiasm in an experiment by telling the subjects what the experiment is about and why you are interested in doing it. This kind of motivation must be used with discretion, however, because it is often better not to let your subjects know too much about the real purpose of some experiments (see the section on *Attitudes and expectancies* which follows).

2. Pay and other rewards. Pay is usually a good motivation, especially if rewards can be set up for extra good performance. In cases where pay cannot be used as a reward (e.g., in using soldiers or sailors in systems experiments), it may be possible to set up other kinds of rewards—time off, being excused from certain kinds of work details, etc.

It is easy to appreciate that pay incentives will affect the outcomes of experiments in which the dependent variable is speed, accuracy, or amount of work done. But these incentives will also affect performance in psychophysical experiments which are generally thought of as being relatively stable in outcome. Blackwell, for example, tested three groups of subjects over a number of days.[13] A control group and an experimental group were tested with a conventional method of constants; another experimental group was tested with a forced-choice method. Between the third and fourth test sessions the two experimental groups were offered some high monetary incentives for better performance. Unfortunately for our purposes here, they were also given a talk by the experimenter on the theoretical importance of the experiment. Thus, the motivation added at this point is complex—we cannot say that it was a monetary incentive pure and simple. Some subjects seemed mostly interested in the financial possibilities; others were concerned about the fictitious theory elaborated by the experimenter. The outcome shown in Figure 32 leaves us with no doubt on one point, however—increased motivation can influence the outcome of psychophysical experiments.

3. Knowledge of results. A very good motivator in many experiments is the simple expedient of telling subjects how they are doing— letting them see their own results. The reason this seems to work is that it puts the subject in competition with other people, and competition is generally a good motivator. Even in the absence of social competition, subjects will usually try to improve their own performance whenever they have some kind of score or measure of their performance. This is the kind of motivation that keeps the amateur athlete trying to better his own record. In experiments extending over many days, e.g., in radar systems evaluations, excellent motivation can frequently be induced by generating a spirit of competition between subgroups of subjects. In all such instances, it is a good idea to post daily results.

Since the effectiveness of information as an incentive is often overlooked, I shall describe in some detail an experiment by Gibbs and Brown.[60] These investigators tested 16 young men, between 17 and 25 years of age, who had volunteered to participate in psychological experiments. Each man was given an uninteresting, unskilled, and repetitive task at which he worked two hours every morning and two hours every afternoon for two weeks. The task was that of operating a document copying machine, i.e., of feeding the machine by hand, processing each item by pressing a foot switch and then removing the item by hand. The material itself was selected to be dull, consisting of separate pages of difficult scientific reports, encyclopedias, and historical reviews. The men were asked to work as quickly as possible, but they were not told to reach or maintain any standard of output. They received a flat weekly payment. No supervision was given during the working session, and, after the essential instructions had been given, the experimenter came into the room only at the beginning and end of each two-hour work period. So far as possible, then, the only motivation operating was that of self-competition and the subject's own satisfaction in working.

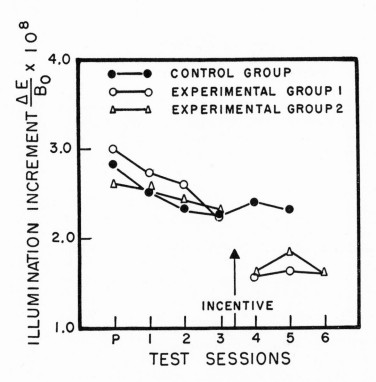

FIG. 32. Average thresholds obtained with three groups of subjects in a psychophysical experiment. P is the practice session; numbers identify the test sessions, each of which was on a different day. The Control Group and Experimental Group 1 were tested with a conventional method of constants. Experimental Group 2 was tested with a forced-choice method. Between test sessions 3 and 4 the two experimental groups were offered high monetary incentives for better performance. (Data of Blackwell[13] from Chapanis.[27])

The experimental variable was provided by a counter which tallied each page as it was processed. In half the trials the counter was clearly visible to the subject; in the other half it was covered. Six subjects worked the first week with the counter visible and the second week with the counter covered; six subjects worked with these conditions reversed in order; four other subjects worked with the counter visible, or obscured, during the entire two-week period. The only function the counter served was that of providing information to the subject about his output.

The results of this experiment show that output was uniformly higher when the men had information about their output. For the men who worked under both conditions, the gain was approximately equal to one day's extra production every four days. The total output of the two men who worked with the counter visible during the entire two-week period was 81,300 items. The output for the two men who worked without seeing the counter was only 37,400 items.

4. Punishment. Although not generally as effective as rewards, punishment can be used for motivational purposes. The punishment may take a variety of forms, e.g., shocks whenever errors are made, having subjects forfeit some time for poor performance, etc.

Perhaps the most important thing to watch for, however, is that in so far as possible *the motivating conditions should be equal for all the subjects in the experiment.* It would not do, for example, to have one group given rewards by way of extra pay; another punished by having free time taken away. The second point to remember is that it is generally better to try to do something positive about controlling motivation rather than to ignore the problem. If you ignore the problem you are merely allowing the subjects to supply their own motivation. This in turn will result in much more variability in motivating conditions than would any experimenter-imposed scheme.

Attitudes and expectancies. Other variables which are extremely difficult to control in human experiments are the attitudes and expectancies of the subjects. The trouble with people is that when they serve as subjects in experiments they have their own ideas about the experiment. If they can see that you want the results to come out a certain way, they may, consciously or unconsciously, help you along. So unless you are careful and take precautions, subjects may bias the results of experiments in the way they think, or think you think, the results *ought* to come out.

Unfortunately, the extent to which subjects may influence experimental results often extends beyond mere biasing. Almost everyone who has worked extensively with people has at some time or other discovered outright cheating or falsification of experimental data. You may impress upon your subjects that this is an experiment to discover new facts, you may instruct them that you want them to do the best they can and that you expect no more, and you may tell them repeatedly that what they do in the experiment will have no bearing upon their jobs or position in life—despite all these precautions, occasionally a person will deliberately give false data in the laboratory. Although such people are rare, they are common enough to constitute a source of potential risk for the human investigator. We have already mentioned this problem in talking about real-life experiments in which the problem of falsification is usually more prominent.

Instructions as controls. One important way—in fact, the most important way—of controlling the subject's attitudes and expectancies in an experiment is by way of instructions. Your instructions are your way of telling the subject what this is all about, what he is to do, and how he is to do it. And what you tell your subjects may affect drastically the results you obtain. For documentation on this point, we can turn to two well-known experiments. In the first of these, three psychologists arranged a simple reaction time experiment.[83] The observers sat in a quiet room and were instructed to release a telegraph key as rapidly as possible in response to a stimulus. In every case the stimulus was a tone. But here was the point: for half of the trials the subjects were instructed that they might expect either a light or a tone and that they should react in the same way to either stimulus; the other half of the trials the subjects were told to expect only a tone. The two kinds of trials were properly counterbalanced to control for practice, fatigue, and other cumulative effects. The important point to note is that the stimuli were the same, the responses were the same, but the "expectancies" differed—in one case, the subjects expected to react to either of two kinds of stimuli; in the other case, they expected to react only to one kind of stimulus.

Average results for 39 college students are shown in Figure 33. The gradual decline in the average reaction time during the experiment can be attributed to learning or practice and need not concern us for the moment. The important point of these data is the difference between the average reaction times in the two curves. When the subjects were prepared to react to either of two stimuli, their reaction times were consistently longer than when they were prepared to react to only one stimulus. In this case, the difference can be attributed to only one variation in what the experimenter did—the instructions he gave his subjects.

The other experiment is on the design of dials.[67] In this study, the investigators used round dials, each of which covered a range of 0 to 100. One kind of dial had 100 marks, one at every unit, and was designated the 100 × 1 dial; another had 20 marks, one at every 5 units, the 100 × 5 dial; the last had marks at every 10 units, the 100 × 10

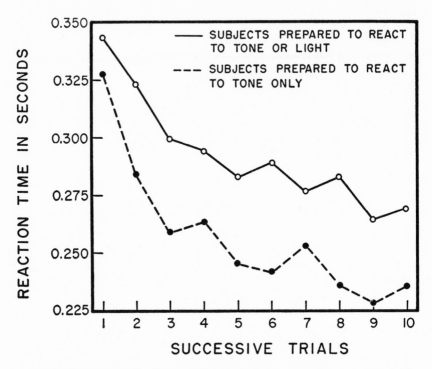

FIG. 33. Average reaction times of 39 subjects to ten successive presentations of a tone. In all cases the stimulus was the same (a tone) and the response was the same (releasing a telegraph key), but the "set" differed. In half the trials the subjects were told to expect only a tone; in the other half of the trials they were told to expect either a tone or a light. (Data of Mowrer, *et al.*,[83] from Chapanis.[27])

dial. Each kind of dial was made in three sizes: 0.7, 1.4, and 2.8 inches in diameter. The men who read these dials were given unlimited time but worked under two different kinds of instructions: (a) speed and accuracy instructions—"Read these dials as fast and as accurately as you can"; (b) accuracy instructions—"Read the dials as accurately as you can."

The results of this experiment are shown in Figures 34 and 35. Although there are many important things we can see about dials from these figures, our interest at the moment is in only one thing—the effects of the instructions. And these, you will have to admit, are great. When the subjects were instructed to be as accurate as possible (Figure 34), they made fewer errors than when they worked under speed and

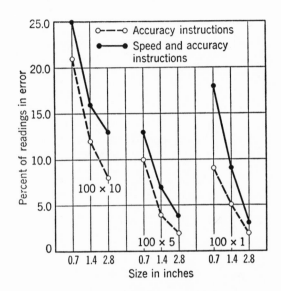

FIG. 34. Errors made by subjects in reading nine different dials. The dials came in three sizes and had three different kinds of markings. Subjects read the dials either under "accuracy" or "speed and accuracy" instructions. Our interest here is on the effect of the instructions. (Data of Kappauf[67] from Chapanis, Garner, and Morgan.[29])

accuracy instructions. The time data in Figure 35 show even more discrepant results. When subjects operated under accuracy instructions, they were very much slower than when they operated under speed-and-accuracy instructions. Remember that these differences were caused essentially by three words " . . . as fast and . . . . "

There are many more experiments we could cite along these same lines, but there is no need to labor the point. The instructions you give your subjects can influence markedly the results you get. *Even if you give no instructions, the subject will set up his own implicit operating rules.*

If you do not tell him whether to be fast or accurate, he'll set up his own working standards, and the trouble is that no two subjects are likely to have the same standards. Thus, if you want some consistency in your data, you had better give instructions.

Some pointers on instructions. Having seen that instructions can influence the results you get in very important ways, we need to con-

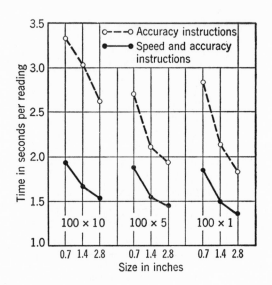

FIG. 35. Average times required by subjects to read nine different dials. These data come from the same experiment from which the data in Figure 34 are taken. (Data of Kappauf[67] from Chapanis, Garner, and Morgan.[29])

sider some rules for making up good instructions. Here are five to keep in mind:

1. Decide first of all exactly what kind of instructions you want to give your subjects. Although this may seem to be an obvious point, it is not. Often, experimenters have not asked themselves, "Just exactly what do I want to tell these people?" Do you want them to strive for accuracy? Speed? Speed and accuracy? Do you want them to do the task in a particular way? Do you want them to figure out their own method of doing the task? These and many similar questions must be answered by the experimenter before he tests his first "guinea pig."

2. Be sure your instructions tell your subjects exactly what you ex-

pect of them. You know what you're trying to find out; the subject who reports for your experiment doesn't. Tell him. Remember that you will probably be asking him to do something he has never done before. Perhaps he is facing a new gadget, the like of which he has never seen before. Tell him what he will see, or hear, or feel. Tell him what he is to do. Tell him how he is to do it.

3. Be sure that your instructions are written simply, clearly, and directly. Don't use high-powered words in explaining the experiment. Remember that you've lived with the experiment for a long time, and you know all about it. You've worked over the apparatus, you've thought about an experimental design, and you've worried about statistical analyses. All this is old stuff to you. But the chances are that your subject will find it all very confusing. So tell him simply. Instead of "This is an experiment on your capacity for pulse-counting 100-per cent modulated illumination intensities," tell him, "We want to see how well you can count light flashes."

4. Be sure your instructions are standard. Write your instructions out and read them to your subjects, without modifications. This is the only way you can be sure to keep conditions constant. If you tell some subjects, "Be as accurate as you can," and others, "Remember that it is absolutely essential for you to concentrate on accuracy and do not make a single mistake," the chances are that the latter group will give you different results from the former. It's difficult for you to say the same words over and over again unless you memorize them, or write them out and read them. If you want your subjects to operate under the same conditions, you must keep your instructions standard.

5. Be sure your instructions are understood. Even after you have worked over your instructions and have translated them into simple, direct English, you are still likely to find that some subjects misinterpret, forget, or just do not understand what it is they are to do. One way of guarding against such possibilities is to have the subjects tell you in their own words what they think they are to do. Although this may do the trick for simple sorts of experiments, it is generally not enough in complicated studies where the subject may have to remember to do a number of different things. About the only way of ironing out these problems in such complicated studies is to schedule a series of practice or trial runs during which the experimenter can observe and prompt the subjects in their tasks.

Other social factors. If you remember that your experiment is a social situation and that there is bound to be interaction between you and your subjects, you will immediately recognize that there are a number of additional variables which need to be controlled. Do you stay in the room or not? Do you correct your subject when he makes a mistake or do you let him go on undisturbed? Do you give him words of encouragement at certain times, or do you remain silent through the experiment?

Fraser, for example, has shown that the presence of the experimenter, even though he may remain out of sight of the subject, will raise the accuracy of subjects in a prolonged visual task resembling that of the inspection process.[55] We could go on and cite still other evidence of this sort, but it will perhaps suffice to say that these and many similar factors must be considered when you work with people. For some kinds of experiments many of these questions are relatively unimportant; for others, they may have very considerable consequences. There is no way of cataloging them all, and the best the experimenter can do is to be aware of the problem and to evaluate constantly the probable effect of his behavior upon that of the subject.

The most important thing to keep in mind, however, is that whatever you decide to do, your treatment of all the subjects should be the same. If you decide to stay in the experimental room, be sure to stay there for *all* trials. If you decide not to stay in the room, be sure to get out for *all* trials. If you decide to offer words of encouragement, do it to all subjects. And so on. Consistency in the experimenter's behavior throughout an experiment is an important form of control which should not be overlooked.

On control observations. Although it is possible to get precise control over many of the important variables in a human experiment, it is difficult or impossible to control some of the most important ones, like attitudes, expectancies, suggestion, and other social factors. To get around this difficulty psychologists frequently collect control observations to protect themselves against the danger of unwarranted conclusions being drawn from experiments in which variables which should have been controlled could not be.

Ask most physical scientists what a control group is, and the chances are you will not get a very good answer. The idea of control observations and control groups, in the sense that psychologists use these terms, is foreign to them. Physical scientists do not need control groups and, in fact, it would be absurd to insist on such controls in most chemical or physical experiments.

The need for control observations: an experiment on fatigue. For a good illustration of the need for control observations in human experiments consider a study by Pincus and Hoagland.[90] On the basis of certain physiological evidence, these investigators had an idea that they could reduce fatigue in factory workers by administering a drug, Δ5 pregnenolone. They went into a factory producing bayonets and selected twelve experienced turret-lathe operators who were making bayonets for army rifles on a rush order. The pay was on a piecework basis, plant morale was high, and there was a war on. Thus there was

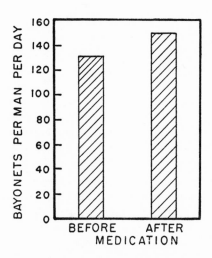

FIG. 36. Average number of bayonets produced per man per day before and after daily administration of Δ5 pregnenolone. The averages are based on the performance of six workers over a four-week period before and after medication. (Data of Pincus and Hoagland[90] from Chapanis.[27])

good reason to suppose that the production incentive was high. Records were taken of a whole month's production, and the average number of bayonets produced per man per day are shown in the first bar of Figure 36. Following this initial trial period, six of the operators were given a daily dose of the drug but were told that this was a "new vitamin type pill" that might lessen their susceptibility to colds. Production records for the month following medication are shown in the second bar in Figure 36 and show an increase of about 14 per cent.

The question is, did the drug produce the result? Let's suppose you were doing a roughly comparable experiment in a steel mill. You have a mixture of melted steel from which you draw off a sample to cool. To the remainder of the mixture you add some chromium and let this batch cool. You find that the latter results in a product which is more resistant to corrosion than the former. Did the chromium do it? The chances are that you would have little question about it. Unfortunately, you cannot let yourself make that kind of an assumption in human experiments.

Pincus and Hoagland were seasoned investigators. So, although they thought they had good conditions for their experiment (trained subjects, high incentive, an adequate sample of production, subjects unaware of the real nature of the drug, and so on), they took the additional precaution of testing a control group. Six of the subjects in this experiment were treated just like the first six, except that instead of getting tablets of pregnenolone, they received tablets of ordinary milk sugar which resembled the drug in taste and appearance. You can see what happened in Figure 37. The control group increased production by almost exactly the same amount as the experimental group! The mere fact of being given a pill—any kind of pill—produced a change in production.

The need for control observations: an experiment on lighting. We have already had occasion to comment on the lighting experiments done at the Western Electric Company when we were discussing preferences as measures of human engineering effectiveness.[95] That same series of experiments illustrates the need for controls even when objective measures of performance are used. In one part of that study a group of operators who wound small induction coils on wooden spools was selected for test. This test group worked under illuminations of 24, 46, and 70 foot-candles, and the production figures for the group seem perfectly clear-cut and reasonable. Increases in illumination were accompanied by increases in production. This is what you would expect on a common sense basis and there seems to be no reason to doubt the results. Or is there?

The catch is that at the time this experiment was started, the experimenter had also selected a control group. The control group matched the test group in experience, and at the start of the tests the two groups had very nearly the same average rate of production. The two groups worked in entirely separate buildings so that the influence of any spirit of competition would be reduced. Although the control group worked under a more or less constant level of 16 to 28 foot-candles—the small variations were produced by variations in the outside illumination—it also showed a marked increase in production and the increase was of almost the same magnitude as the changes shown by the test group!

This kind of experimental result could be illustrated over and over again. Sometimes the results have been merely wrong. In some instances, however, they have been dangerously wrong. The results of human experiments do not always mean what they seem to mean.

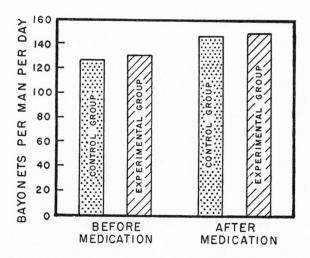

FIG. 37. The experimental group in this illustration is the same as that shown in Figure 36. The other data are for a control group which matched the experimental group except that the workers in the control group received every day a milk-sugar pill which looked like the pill containing the drug. (Data of Pincus and Hoagland[90] from Chapanis.[27])

Methods of collecting control observations. The general principle governing the use of control observations is that *control observations are collected on subjects who are treated in exactly the same way as experimental subjects except for the critical aspect of the experiment.* In the Pincus-Hoagland experiment, the control group was treated like the experimental group in every respect *except* that the pills administered to the former did not contain the critical ingredient. In testing the training effectiveness of a training device, the psychologist would always use a control group which received exactly the same kind of treatment as an experimental group *except* for practice on the device in question.

There are four different ways of collecting control observations.

Three of them involve using a control group which is different from the experimental group. The fourth involves collecting control observations on the same subjects as are used for the experimental observations. These four ways of collecting control observations are, in fact, the same as the four methods described on pages 157 to 160 for making simple comparisons. They are:

1. The use of independent groups.
2. The use of matched groups.
3. The use of matched subjects.
4. Having each subject serve as his own control.

All of the considerations about the selection of subjects, experimental design and statistical analysis which were discussed on pages 157 to 165 are relevant here. You should now read that section again with this new application in mind.

CONSULTING THE LITERATURE

At about the same time the experimenter is defining the question he wants to answer, he should consult the literature to see what relevant experiments have already been done in the area. There are a number of reasons why this is a worthwhile step.

The experiment may already have been done. Psychologists have been in the business for quite a while, and it often comes as quite a surprise to people outside the field to discover the number and variety of applied problems they have investigated. In some areas of experimental psychology, you are on safe grounds if you go ahead on the assumption that someone thought of your experiment before.

Related researches may yield valuable ideas about ways of doing your experiment. Even though the precise problem you want to investigate has never been done before, the chances are good that you will get many valuable ideas from experiments which resemble or approximate the one you want to do. In particular you may find something helpful about the following:

1. Apparatus. Photographing a pilot's eye movements in flight, measuring training with a radar simulator, measuring pursuit tracking, and any of a number of other human engineering experiments require

elaborate apparatus about which you can get good ideas from other experiments.

2. The number and kind of subjects. Some experiments require a lot of subjects; others can give good results with only a few. Some require special kinds of subjects or require that certain kinds of subjects (for example, color-blind ones or tone-deaf ones) be excluded. Consulting the literature will often give you some clues about what you might have to plan for.

3. Relevant variables. It is almost impossible to think of all the relevant variables in an experimental situation. Related experiments may often suggest important factors that you had not originally considered.

4. Mistakes to avoid. Reading the literature may help you see the pitfalls that other experimenters have fallen into and suggest ways of avoiding them.

Where to go for information. A good place to start looking for information is in the Tufts *Handbook*.[107] It contains a selected bibliography of several hundred items of great use to people in the human engineering area. Another valuable source is the bibliography by McCollom and Chapanis which contains over 5,000 references classified under 94 headings.[77] The psychologist's standby is the journal *Psychological Abstracts*, now in its thirty-third year of publication. This journal can be found in most university libraries or in the private libraries of most professional psychologists. In it are published each year thousands of abstracts of studies on psychological topics. Each item is classified and indexed under appropriate headings.

Two other periodicals which are of direct interest to the human engineer are *Ergonomics*, a British journal devoted to human factors in work, machine control and equipment design, and *Human Factors*, the journal of the Human Factors Society. Browsing through the back numbers of these two publications is almost certain to be rewarding to the investigator in this field.

WHO WILL BE YOUR SUBJECTS?

The next general consideration the experimenter faces in setting up an experiment is the selection of experimental subjects. The primary trouble here is that people differ. In fact, there is no human trait or characteristic in which people do not differ significantly.

THE NUMBER OF SUBJECTS

The first important consequence of individual differences is that you should plan on an adequate number of subjects. This, perhaps, is one of the most serious errors which physical scientists and engineers make when they venture into fields of human experimentation.

As a matter of fact, psychologists themselves are not free of this criticism. One of the most critical difficulties one encounters in applying the results of many classical psychophysical studies is that they were done on only one subject.[24] In human experimentation there is no reliable way of deciding in advance what constitutes a "typical" or "average" subject. The best you can hope to do is to test *enough* subjects so that you can get a dependable measure of average performance and some estimate of the amount of variability you can expect to find.

Consider, for example, the rather smooth and uniform learning curve in the middle section of Figure 29. This curve is smooth primarily because there are eight subjects contributing to it. Figure 38 shows the learning curves for five of these eight subjects and you will have to agree that individually they appear to be quite erratic. Not only are there marked differences between the mean *levels* of these curves, but their *shapes* are different. Subject *A* shows the biggest drop from beginning to end, with a marked decrease between the fifth and sixth days. For a period of four days (3 through 6), Subject *C* actually got worse from day to day. Subject *H* shows the least learning over the entire eight-day period with a smooth transition from day to day.

A much more impressive example occurs in the study on the effects of anoxia reported in Table 15 (Chapter 4). In that study each subject read a visual test chart three times: the first time with nearly normal oxygen concentrations (at a simulated altitude of 2,000 feet); the second time under conditions of mild anoxia (at a simulated altitude of 15,000 feet); and the third time with an adequate oxygen supply (at 15,000 feet breathing 100 per cent oxygen). The average data for twenty subjects show that fewer test symbols could be read under the anoxic conditions than could be read under the other two test conditions. You will recall too that the statistical analysis of these data showed the differences between the averages to be highly significant.

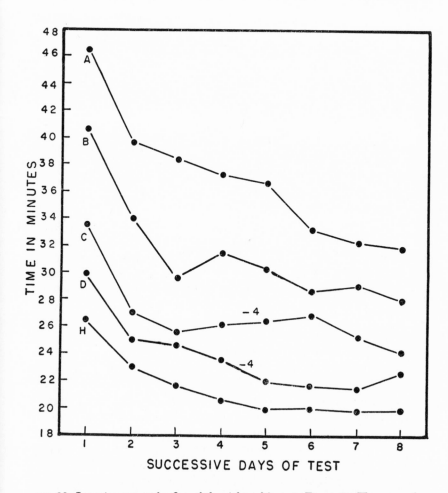

FIG. 38. Learning curves for five of the eight subjects in Figure 29. The curves for subjects C and D have been moved downward by four minutes; the curves for the other subjects are correctly positioned with respect to the scale on the left. Although the average curve in the middle section of Figure 29 is fairly smooth, individual learning curves show marked irregularities. (Unpublished data from an experiment by Scales and Chapanis[97] from Chapanis.[27])

Although these *average* data are quite dependable, note what a large amount of variation there is among individual subjects. Two subjects, in fact, gave data completely contrary to the average trend since they saw *more* test objects correct under the anoxic conditions than under the other two conditions. You can readily appreciate that if you had run this experiment with only two subjects and had happened to get *A* and *M* you might have been completely misled about effects of anoxia on this visual task.

Unfortunately it is very difficult to forecast in advance how many subjects one should plan for in an experiment. Most experimental psychologists acquire some intuitive notions about the numbers of subjects that should be used in different kinds of experiments, but even they are frequently wrong. As a general rule, however, psychophysical experiments on simple sensory functions (e.g., the detectability of radar signals on radar scopes, the visibility of signal lights, the masking effects of noise, etc.) generally require fewer subjects than most other kinds. Learning experiments, experiments involving motor performance, and so on, may give erratic results unless a minimum of about twenty or thirty subjects is used. In any event, it will be a rare human engineering experiment that will give you definitive results with only two or three subjects.

HUMAN SAMPLING PROBLEMS

The second consequence of human variability is that any human experimentation inevitably means sampling. It is impossible to test, or question, an entire population, even with the resources of the federal government. The national census comes close to being a complete human survey, but the Census Department has to admit that even its counts are not complete. People move, die, get born, or change in other respects during the time it takes to make a census. Thus, in general, one must always be content with a sample. And the selection of this sample is what causes most trouble. How do you pick the people on whom you will do your experiments? If you use men, your results may not necessarily apply to women. If you pick college students, your data might not hold for the average high-school graduate. Results obtained on young adults are almost certainly not true of teen-agers or octogenarians.

These remarks may seem, at first glance, to contradict something I said in Chapter 5. In discussing the problem of realism in human engineering experiments (pages 198 to 207) I pointed out that comparisons of alternative equipments are very often stable even when they are tested with different kinds of subjects. Although this reassuring finding has turned up in several studies, it is not universally true. Then too, many studies try to end up with more information than the simple statement "equipment A is better than B." The experimenter may want to know, for example, "About what percentage of errors may we expect to find if operators use equipment A?" The answer to this question, which requires a precise numerical result, is much more sensitive to the particular sample of subjects tested than is the outcome of a simple comparison.

In any case the careful experimenter always gives some thought to the selection of his subjects. Since entire books have been written on the problem of human sampling,[40, 118] we can do little more than to point to this problem, emphasize its importance, and discuss a few solutions.

Fortuitous sampling. The most common type of sampling scheme used in human experiments is one we may call fortuitous sampling: The experimenter prevails upon a few of his friends to assist him by serving as subjects. A serious criticism of many human engineering studies is that engineers serve as their own subjects. The trouble with this is that engineers are atypical—they are more intelligent than average, they have a better than average familiarity with numbers, they understand complicated pieces of machinery, and so on. A much more serious problem is encountered when engineers who have been associated with a particular development serve as subjects in the evaluation of it. No matter how much they may try to be fair, their "ego involvement" in the machine system undoubtedly influences their performance even in such simple matters as speed and accuracy. Many pieces of military equipment which are poorly human engineered are probably that way because they were tested only on engineers.

Psychologists are not free of this criticism, either. Unfortunately, many psychological experiments are done on undergraduate college students, primarily because students are so handy. This state of affairs has led one psychologist to remark—only half in jest—that it might be more appropriate to rename some of our textbooks, *The Psychology of the American College Sophomore.*

Random sampling. Most statistical tests of significance are derived on the assumption that subjects in the sample are randomly selected from some population. By random we mean that every member of the population has an equal probability of appearing in the sample. By population we mean all those individuals who have certain common characteristics. Practically speaking, we may define a population in any way which seems reasonable for the problem at hand. Here are some sample populations:

1. All able-bodied males in the United States between the ages of 18 and 35 on July 1, 1959. (This implies, of course, that we have specified what we mean by "able-bodied.")
2. All rated Air Force pilots between the ages of 25 and 50 who have had at least 1,000 hours of flight time.
3. All naval enlisted personnel with IQ's above 130.
4. All color-blind males in the United States between the ages of 20 and 60.
5. All army recruits between the ages of 18 and 30 who are at least 6 feet tall.
6. All female engineers employed in American industries on January 1, 1959.
7. All graduates of the Fleet Sonar School during the period 1956–59, inclusive.

To show how you would pick a random sample, let us suppose that you want to test two different pieces of sonar equipment. For these tests you decide to use 20 sonarmen picked from population number 7 above. Start by assigning the numbers 1 to N to each man who fits the description. Now use a table of random numbers to pick out 20 consecutive numbers. These numbers identify the men who will be your subjects.

Although psychologists use random sample statistics all the time in describing and interpreting their data, practically never will you find that their subjects were selected in this way. For most purposes this discrepancy between theory and practice is probably not as serious as it sounds. Nonetheless, it is a shortcoming to keep in mind particularly in interpreting data which have only marginal statistical significance.

Stratified sampling. A sampling procedure which has merit for many kinds of experiments is a stratified sampling procedure. For example, in a study of keyset design Lutz and Chapanis were trying to find out where people expect to find numbers and letters on certain configurations of keys.[76] Since they were not sure whether men would give different results from women, they decided to use 50 men and 50 women.

They further divided each of the sex groups into three age groups. Finally, since they suspected that familiarity with keysets (such as those on typewriters, calculators, and musical instruments) might be an important consideration, they further subdivided the six age-sex groups into naïve and experienced sub-groups. The complete definition of the sample is shown in Table 36.

As it turned out, for most of the data there were no significant differences between men and women, between the various age groups, or between the naïve and experienced subjects. The advantage of such a sample is that it helps the experimenter to find differences if there are any. If no differences show up, the experimenter has increased confi-

TABLE 36. *In a study of keyset design the group of 100 subjects was composed of sampling sub-groups containing the frequencies shown here (from Lutz and Chapanis[76])*

	Men			Women		
Age group	20–30	30–40	Above 40	20–30	30–40	Above 40
Naïve	8	9	8	8	9	8
Experienced	8	9	8	8	9	8

dence in the generality of his findings. Most of the findings of the Lutz-Chapanis study apply to both men and women, all age groups, and for people who have previous experience with keysets as well as those who have not. This kind of generality is important in itself.

Proportionate stratified sampling. Another successful, but more difficult, way of dealing with the knotty problem of human sampling is to use a proportionate stratified sample. In essence this means that the number of subjects in each sub-group of the sample is proportional to the number of people in each corresponding sub-group of the population. Let us suppose, for example, that we are concerned with getting a good estimate of the average reaction time of automobile drivers. Let us say that the proportion of male to female drivers in the United States is 60–40. In our sample, then, we would want to use 60 per cent men and 40 per cent women. Age is certainly an important determinant of reaction time, so we would want to select men and women of specified age groups. The numbers in each group would again be proportional to the age distribution of men and women drivers. Finally, we might also want to classify people according to other criteria, for example, years of driving experience. A hypothetical stratified sample according to these criteria is shown in Table 37. There you can see that for this experiment we must test exactly 19 men who are between the ages of 20 and 30 and who have had less than one year's driving experience. We have to test 17 women between the ages of 30 and 40 with more than one year's driving experience, and so on. There are many

TABLE 37. *A hypothetical proportionate stratified sample of subjects for an experiment on the measurement of the reaction times of automobile drivers (from Chapanis[27])*

Driving experience	Male		Female	
	Less than one year	More than one year	Less than one year	More than one year
Age				
20–30	19	17	15	7
30–40	22	26	17	17
40–50	8	14	6	10
50–60	2	12	1	7

special technical points to be considered in using sampling schemes like this, but this illustration will suffice to show what is possible in this field.

What characteristics do you sample? Having decided on a sampling scheme—that is, on how to pick your sample—you have to decide *what* to sample. The exact human characteristics you sample will depend greatly on your problem. If you are concerned with a human engineering problem relating to aircraft controls you will undoubtedly want to sample characteristics which are reasonable from the standpoint of the ultimate users of the equipment—pilots. If you are concerned with the design of secretarial chairs you should obviously be sampling other sorts of characteristics.

The list which follows is only a reminder of *some* of the characteristics which can be sampled:

1. General descriptive characteristics, e.g., age, sex, origin (i.e., whether urban or rural).
2. Physical characteristics.
   a. Body dimensions, e.g., height, weight, arm reach.
   b. Sensory characteristics, e.g., visual acuity, auditory acuity, color vision.
   c. Psychomotor characteristics, e.g., strength, reaction time, handedness.
3. Intellectual characteristics.
   a. General intelligence.
   b. Specific aptitudes, e.g., verbal ability, numerical ability, mechanical aptitude.
4. Personality and attitudinal characteristics, e.g., co-operativeness, interests, leadership ability.
5. Experience.
   a. General education, e.g., grammar, high school, or college.
   b. Specialized technical education, e.g., pilot training, engineering training, sonar training.
   c. Specialized experience, e.g., combat experience, certain types of work experience.

Long term changes in the population. Unfortunately for us, people are not only unstable from moment to moment, and day to day, but there is good evidence that the basic characteristics of the population are themselves undergoing long-term changes. What this means, in effect, is that certain human engineering experiments done twenty years ago might not be valid today.

It is not hard to find practical illustrations of this point. One thing the engineering psychologist is usually concerned about is that machines fit their human operators. However, we know that modern man is taller than his grandpa, and that he seems to be increasing in average height over the years. Although this is a very slow and scarcely noticeable change, it is appreciable enough so that the American Standards Association has recently revised its safety code for the protection of operators of power machinery. Earlier editions of the code required that transmission equipment be guarded to minimum heights ranging from 5′–6″ to 6′–6″. This is no longer good enough for today's worker, and minimum codes now extend to 7′–0″.

There are many other respects in which people today differ from their parents and it should be easy for you to see how this might affect the design of machinery. The average age of the population is increasing. Due to medical advances, people live longer, and this means that we have a greater proportion of old workers than we had twenty or thirty years ago. People today are better educated, are more familiar with scientific terms and with gadgets than their parents were, and so on. All of these long-term changes mean that we must continually re-evaluate our principles and results in terms of our present population.

The use of naïve subjects. A general rule of importance in the selection of subjects is that, if you want a fair comparison of alternative human engineering designs, you should generally use naïve subjects, that is, subjects who have had little or no experience with any of the equipments under test. If you use subjects who are experienced with one or more of the equipments you are likely to run into what psychologists call *habit interference.* A case in point concerns the use of simplified typewriter keyboards. Extensive experiments have shown beyond any reasonable doubt that it is possible to make substantial improvements on the present standard typewriter keyboard. But if you were to test such an improved design with experienced typists you would find that the new design is markedly inferior, primarily because the typists have to *unlearn* a set of well-established habits and learn an entirely new set. Of course, if you continued the experiment long enough the superiority of the improved design would eventually show up, but the completion of such an experiment would run into months of continued testing for each subject.

It takes about six weeks of practice for the average typist to acquire some proficiency, and if she has been trained on a standard keyboard all of this training has to be replaced with new training. In general, a person who has to unlearn a set of old habits and then learn a conflicting set of new habits requires much longer for the learning process than a person who learns the new set of habits from scratch. On the other hand, the relative merits of different keyboard designs will start to show up almost immediately with naïve subjects, that is, subjects who have never used a typewriter.

As another illustration of the same principle, Gardner and Lacey were interested in comparing several different types of attitude indicators for aircraft.[57] These indicators were something like those in Figure 1. Two basically different design principles were tested in this experiment: the kinds of movements that are shown in a standard instrument; and movements which were exactly reversed from the standard (the experimental indicator). Two groups of subjects were tested in this experiment: experienced pilots; and naïve college students. The pilots, of course, had had extensive experience with the standard instrument, so it was not surprising to find that they made more errors with the experimental indicator. When the same instruments were tested with subjects who had never used either type of indicator, the results were in favor of the experimental indicator.

Another reason for using naïve subjects in comparing different equipment is that operators who have spent years in using a particular piece of machinery, or a certain work arrangement, will frequently resist changes even though the changes are obviously for the better. Experienced operators sometimes develop a sort of emotional or attitudinal block and stubbornly refuse to admit that the newer equipment or arrangement has any merit. If subjects approach an experiment with such attitudes, it is obviously very difficult to get a fair comparison of the inherent advantages of alternative ways of doing things. Naïve subjects, on the other hand, are less likely to approach an experiment with such biases.

In industry it is sometimes almost impossible to do a genuine experiment using seasoned workers. One first of all runs into complex problems of motivation. If the workers get the idea that the new equipment may be better and so result in someone's being fired, they will almost certainly not give the new equipment a fair trial. There may be other complex kinds of motivation as well. In addition, the influence of labor unions is so strong in many industries that it is impossible to get honest co-operation from everyone concerned in evaluating alternative equipments. Case studies by Roy[96] and Dunlap[43] show clearly how the advantages of technological improvements and human engineering changes may be more than offset by "slow downs" and deception. In such situations, the only way of getting valid data may be to use subjects who have no ego-involvement in the device or procedure, i.e., to use naïve or disinterested subjects.

Above all, never use subjects who have designed the equipment or who have any stake in its potential usefulness. Such subjects, which include design engineers, office personnel, supervisors, and others concerned with a project, are usually not typical in training, experience, sex, and other characteristics of those who will operate the equipment in actual use. Even if they were typical, they must be disqualified because of their "ego involvement" in the outcome of the tests. A person who wants one system to be better than another, or expects it to be, is prejudiced. No matter how hard he tries to be fair, his prejudices influence his performance even in such simple matters as speed and accuracy. Those connected with equipment know too much about it. They probably understand it better than all but a few of the best-trained personnel who will operate it. As a result, they know how to compensate for its weaknesses and to make it show up better than it will under field conditions.

Throughout this section we have assumed, of course, that the experimenter is interested in getting an appraisal of the inherent merits of alternative equipment designs. In applying his findings, the human engineer would obviously have to consider the amount of retraining which would be involved if the newer equipment were installed in some operating situation. Answering this question would require the use of subjects who were already trained in a conventional, or standard, method. The important point to remember is that this is a different question one might ask and that the subjects should be selected accordingly.

The volunteer subject. The final point about subjects is that volunteer subjects are frequently atypical. There has begun to accumulate a fair amount of information which seems to show that people who volunteer for human experiments are different from those who do not volunteer.[16, 69, 94] The differences seem to be primarily in personality and motivation, although other differences have been reported as well. The evidence so far suggests that the general use of volunteers is likely to bias the results of only a few restricted classes of human engineering studies: Those which attempt to describe the "typical" or "normal" reaction to stress, fatigue, or drugs.

APPARATUS

As a group, engineers and physical scientists know much more about the design of particular kinds of equipment than psychologists do. For this reason we shall talk about experimental apparatus for human engineering studies only in the most general terms.

Apparatus should measure accurately and consistently. The first general requirement of apparatus for human engineering studies is that the apparatus should measure accurately and consistently within the range of values where the human stimuli and responses are likely to occur. In this connection it may be worth pointing out that complex apparatus is not necessarily as good as engineers sometimes think. Equipments, especially complicated equipments involving many mechanical and electronic components, may vary enormously from each other and from day to day. In one large unpublished study from The Johns Hopkins University, seven variables were studied systematically to find their relative contributions to the variability of radar readings. These variables included:

1. Differences between experienced radar operators.
2. Differences within operators, i.e., the differences between readings made by the same operator on the same targets on different occasions.
3. Differences between radars on identical targets.
4. Differences in radars during the day as a result of (a) having been turned off overnight or (b) having been left in "standby" condition overnight.
5. Differences between morning and afternoon readings on the same radars under otherwise identical conditions.
6. Day to day differences in the radars.
7. Differences attributable to the radar simulator.

The outcome of the experiment showed that operators contributed a negligible fraction of the variation in target readings in the experiment. By far the greater share of the variability came from the machine element in the system, and the largest single source of variation came from the condition of the radar in the morning, i.e., whether it was (a) cold or (b) warm because of having been left in "standby" all night. Variations between radars were much larger than differences between operators.

Figure 39 shows calibration errors in a particular kind of radar on two successive days. In this experiment, actual tests with radar operators were never begun on any day until the radar technicians said that the equipment was calibrated and ready to go. Figure 40 shows essentially the same sort of calibration errors in bearing-readings between two identical radars in two parts of the same experiment. Part of the difficulty here is that the calibration procedures established by the manufacturer for the equipment were essentially internal electrical calibrations, which are not necessarily related to the consistency with which targets will appear on the face of the PPI.

In his studies of detectability on radar scopes, Williams found so much variability in cathode-ray tubes that he could not use electrical measurements of circuit parameters as the independent variables.[113] He had to use a psychophysical criterion instead. Thus, for example, he identifies *bias* "as a voltmeter reading taken from a visual reference as

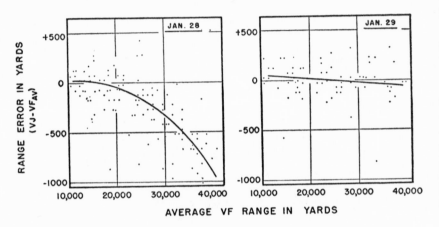

FIG. 39. Range errors in a *VJ* radar as a function of the range of the target on two consecutive days. The data on the left exhibit extremely large calibration errors which were undetected during normal check-out. (From Chapanis.[23])

zero." Not only is there great variability in circuitry, but tubes themselves differ when they are new, and they change as a function of their age.

The experience with radar equipment cited above could be amplified with still other kinds of equipment. Perhaps the most important point to remember here is that when complicated equipments are being used in human engineering studies, it is well to check on the equipment to be sure it is as accurate and consistent as you think it is, or should be.

Equipment sampling. An important consequence of the variability among equipments is this: In many human engineering studies the problem of equipment sampling is every bit as important as the problem of human sampling about which we spoke earlier. Aircraft of the

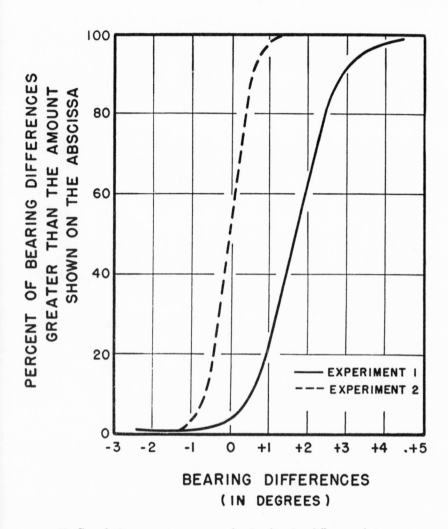

FIG. 40. Cumulative percentage curves showing bearing differences between two *VF* radars in two consecutive experiments. (From Chapanis.[23])

same model differ from each other, and pilots even get to characterize some planes as "mushy"; others as "touchy." The clutch and brake pedals on some automobiles are tight; on others loose. Similar sorts of variability can be found among tanks, training devices, and a great variety of machines. It is not hard to find reasons for such variability. The precision with which the equipment was originally made and assembled, the amount of wear and tear to which it has been subjected, the adequacy of the maintenance—all these contribute to making certain kinds of machines as individual and distinctive as people.

Whatever the causes, the results raise some vexing problems for human engineering studies. Suppose that you had just completed an experiment to find the tracking constants which characterize the control system for a helicopter. Let us suppose that you had done the experiment with a brand-new helicopter fresh from the assembly line. How valid are your findings for a helicopter which has been in service for a year and has been bounced on the deck of an aircraft carrier a dozen times? We have no ready-made answers to this kind of question, because the problem seems to have been largely ignored. Nonetheless, it is a problem to which the investigator should be alert.

One practical approach is to design multi-variable experiments whenever possible and to use two or more different machines of the same kind as one of the variables. The machines should be deliberately selected to cover the range of variability one might expect to find in actual usage. When the results are in, an analysis of the interactions between the machines and the other experimental variables will help you decide whether machine variability is a significant factor.

The apparatus should not interfere with, obscure, or change the behavior under investigation. Many of the equipments used in human engineering studies are, unfortunately, complicated and formidable. Nonetheless, a good rule to keep in mind is that the equipment should not interfere with the behavior under test. There are two ways in which this could happen: (1) the apparatus may be so formidable that it might actually scare the subject; (2) the equipment might actually interfere with the action under study.

The human centrifuge is an example of a device which has an adverse effect on subjects because it is so large and because it obviously spins subjects around at a terrifying rate. There is little one can do to simplify so large a device and, in the face of this difficulty, it is well to give the subject a period of familiarization on the device before actual experimental tests are begun. The period of familiarization may, of course, have to extend over several days or even weeks.

Devices which interfere with the action under study are not hard to find. Studies have been done on the eye movements of pilots during flight. Almost any device for recording eye movements imposes some restrictions on the pilot. Insofar as possible, however, these restrictions should be minimized.

The apparatus should provide means of collecting data. We have already commented on the unreliability of the human observer. One of the positive things one can do to minimize this source of variation is to construct apparatus which provides, automatically if possible, an objective, permanent record of the behavior of the subject. Error counters, error integrators, timing devices, photographic records, magnetic tape recordings, and so on, are some of the ways of collecting objective data on the performance of the subject. They should be used whenever possible.

The apparatus should provide for adequate situational sampling. Earlier we spoke about sampling problems in the design of human engineering studies, but our emphasis at that time was on the problem of sampling people. There is another kind of sampling problem which is equally important in certain kinds of human engineering studies. This, for want of a better term, we may call situational sampling. The apparatus should provide for an adequate sample of conditions or situations.

For example, suppose that you are doing a study aimed at evaluating two radars. If the radar simulator is capable of displaying only a half dozen targets at certain fixed locations, this will not provide a very good test of the radar. So small a number of targets does not provide a good sample of the capabilities of the radar. Similarly, if there is only one small (canned) course available in a tracking device for testing different types of controls, this is not as good as having several different courses. One course may not be an adequate sample of the kinds of situations in which the control might have to be used.

The apparatus should provide stimulus conditions which are varied enough so that they cannot be learned. A second reason for providing an adequate sample of conditions or situations in the apparatus is to avoid the possibility that subjects will thoroughly learn the peculiarities of the apparatus and so not give you data about the thing you are interested in. Suppose that we wanted to compare the readability of two scales on a radar scope and that we had a total of a half dozen targets to do it with. You can readily see that if operators were tested over and over with these targets they would soon recognize each one and start giving the same answers because they knew they were the same.

Thus, our test might not reveal any difference between the two scales even though there really was one. In learning the peculiarities of the equipment subjects may give stereotyped answers which might have nothing to do with readability, but only with memory.

There is a good illustration of the difficulties one can get into on this score in a study reported by Faber.[47] This investigator was interested in the accuracy with which pilots could track an erratically moving target as a function of the stick force and displacement. The motion of the target was produced by a cam driven at one revolution per minute. The target motion was the sum of the first 24 harmonics of a sine curve. The harmonics were of equal amplitude but had random phase relationships. It was an erratic movement and, of course, it did not repeat itself for 60 seconds. The standard procedure was to test subjects under a number of conditions and, for each condition, to have the subject practice for 4 minutes and make a run for the record of a minute's duration. Although the cam was run backwards and forwards, the investigator soon discovered that subjects were learning the target motion. Thus, it was necessary to make a second cam for the subjects to practice on. You can readily see some of the troubles this can create. Are the results really a reflection of the control variables, or are they the results of learning? If the various experimental conditions were tested in a fixed sequence for all subjects, instead of being randomized as we recommended earlier, we may have some serious question about whether the data constitute anything more than learning curves.

This problem shows up in a great many other ways. For example, radar instructors become so familiar with the terrain over which they fly day after day that they may often be useless for running any comparative tests on radar modifications. As another example, an automobile manufacturer was testing the effects of tinted windshields on visibility. He arranged a driving course with man-sized targets which would pop up at various points along the way. Test drivers were required to state when they first saw each of the target figures, and the average distances at which the figures were visible constitute the data. The results show that the tinted windshields had no effect on visibility —at least that seemed to be the result. A psychologist who looked at the experiment, reviewed the data, and talked to the test drivers soon discovered that the drivers had early memorized the locations of the targets. Do you think that the results of this study are to be trusted?

| Chapter 7 | *The psychophysical methods* |

The psychophysical methods are special experimental procedures that are used time and again in human engineering studies. Psychophysics originated in the nineteenth century as a science of the functional relations between "body and mind." In the years that followed there was built upon these basic methods a considerable amount of theoretical superstructure about the "measurement of sensations," "body-mind relationships," and so on, much of which was subsequently shown to be trivial or based on faulty grounds. Since arguments about some of these problems still persist in learned psychological circles, it is useful to distinguish between the theoretical icing and the solid cake which supports it. As basic tools for investigating a variety of practical problems, the psychophysical methods are sound and useful. Perhaps the best evidence for this is that they have been widely adopted into many phases of engineering—telephone, radar, illuminating and geophysical engineering being some good examples. In this section we shall be entirely concerned with the psychophysical methods as tools for certain kinds of man-machine investigations.

DEFINING THE PSYCHOPHYSICAL METHODS

Just looking at the word *psychophysical* gives you a clue about what it means. It is made up of two parts, *psycho-* and *-physical*, and thus we may say broadly that these methods are concerned with the measurement of relations between psychological magnitudes and physical magnitudes. Let us look at this idea in greater detail.

Physical versus psychological dimensions. Many physical dimensions are correlated with psychological ones. For example, the intensity of radiant energy (between the limits of about 400 and 750 millimicrons or 4,000 and 7,500 angstrom units) is correlated with brightness, the wave length of monochromatic radiation is correlated with color, the intensity of acoustic energy is correlated with loudness, the frequency of acoustic energy is correlated with pitch, and so on. In these illustrations, *intensity, wave length,* and *frequency,* refer to physical magnitudes or measurements, while *brightness, color, loudness,* and *pitch* refer to psychological magnitudes. By *physical* magnitudes we mean quantities that can be measured in the laboratory with physical measuring devices: rulers, scales, meters, photocells, and the like. By *psychological* magnitudes we mean the amounts of sensations we experience as "mental" phenomena. The latter cannot be measured with simple physical instruments.

In psychophysics it is customary to refer to a particular value of a physical dimension as a *stimulus.* The corresponding psychological term is *sensation.* Although it is difficult to find two psychologists who will agree on how to define a sensation, for our purposes we may say that a sensation is the experience which results when a sense-organ is excited by a stimulus. For example, monochromatic radiation of 525 millimicrons is a stimulus. What we see when we look at such a stimulus is the sensation *green.* Water at a temperature of 55°F is a stimulus. What we feel when we step into a bath of such water is the sensation of *coldness.*

The general purpose of the psychophysical methods. As we have already indicated, physical and psychological measurements go together, and psychological events are often a monotonic function of the physical ones. In general, the greater the intensity of radiant energy, the brighter the sensation it produces. But for many theoretical and practical purposes it would be useful to have some common reference points in the two systems of measurement. This is what the psychophysical methods help us to get.

Although the psychophysical methods were originally developed for, and have had their greatest usefulness in, studying the relationships between stimuli and sensations, they are now being adapted to the study of much broader and more complex human problems—the measurement of attitudes, mental abilities, the naturalness of the

voice in telephonic transmissions, and so on. Some of these more complex problems have great human engineering interest, but in this
chapter we shall confine ourselves to the simpler applications.

KINDS OF PROBLEMS WHICH MAKE USE OF PSYCHOPHYSICAL METHODS

Textbooks on psychology identify seven or more different classes of
problems that can be investigated by the conventional psychophysical
methods.[62] Of these, three are of more engineering interest than the
others. They are:

1. The measurement of absolute thresholds: What value of a stimulus marks the boundary between sensation and no sensation?
2. The measurement of differential (or difference) thresholds: What
   is the smallest detectable *change* in the value of a stimulus?
3. The measurement of equality: What values of two different kinds
   of stimuli produce the same psychological effect?

Absolute thresholds. It is always possible to find some value of a
stimulus which has no effect on human behavior. You can always find a
light intensity so low that no one can see it. You can find a movement
so small that no one can detect it. A *threshold* (or *limen* as it is sometimes called) is the value which divides the range of stimuli into two
classes: (1) those stimuli which the human can sense lie above the
threshold (and are called *supra-threshold* or *supra-liminal*); (2) those
which he cannot sense lie below it (and are called *sub-threshold* or
subliminal).

It is easy to find illustrations of problems which involve the measurement of absolute thresholds. What is the faintest light signal you
can see on a dark night? What is the faintest spot you can see on a
radar scope? What is the faintest burst of acoustic energy you can
hear? And so on.

Differential thresholds. Just as you can always find some value of a
stimulus which is so low that it is imperceptible, so it is possible to
change a stimulus by such a small amount you cannot detect that any
change has occurred. If you are carrying a 100-pound load you will not
notice it if someone adds a feather to your burden. If you are standing
near an air hammer you may not hear the additional sound created by
someone talking.

A practical problem involving differential thresholds has come out of transmission research in the Bell Telephone Laboratories. Although it is a telephone research organization, BTL is concerned with television too, because telephone wires and circuits carry TV programs from one end of the country to the other. There are a large number of design factors that influence the sharpness of TV pictures which are transmitted over telephone lines—bandwidth and noise being just two. How much can you restrict the bandwidth of the circuits before you produce a noticeable change in the clarity of the picture? Or, how much noise will produce such a change? Both of these questions are practical ones involving the measurement of a differential threshold—a complicated one, to be sure—but a differential threshold nonetheless.

The *differential threshold* is often called the *difference limen* and is abbreviated *DL*. In addition you will often find the symbol Δ used in connection with difference thresholds. Thus, Δ*I*, for example, usually means a just-noticeable change in intensity, that is, the differential threshold, or difference limen, for intensity; Δ*f* usually means a just-noticeable change in frequency; Δλ a just-noticeable change in wavelength. And so on.

The measurement of equality. Qualitatively different stimuli may produce the same effect in so far as some psychological dimension is concerned. Imagine the following simple demonstration: Project a red light onto a section of screen and next to it project a green light. By putting enough neutral filters in front of the green light you can cut its intensity down until it is obviously darker, that is, less *bright*, than the red light. The two lights still differ in color, but they also differ in *brightness*. You can also reverse the procedure—by putting neutral filters in front of the red light you can also cut its intensity down until it looks darker than the green light. By proper juggling of the neutral filters in front of the two lights you will come to a point where you cannot say that one is brighter or darker than the other. Although they *differ in color*, they now *match in brightness*.

Careful experiments of this sort have been done throughout the spectrum to give us the luminosity function—a curve which tells us the amount of radiant energy required at different wave lengths to produce colors of the same brightness. This function is fundamental to our entire system of heterochromatic photometry,* and is an important basic concept in illuminating engineering. Similar sorts of measure-

* Heterochromatic photometry is the photometric comparison of light sources which differ in color.

ments have been carried out in acoustics to give us equal-loudness functions. The important point to note about these illustrations is that they are functions relating physical magnitudes to psychological ones, and that the essential kind of measurement is a measurement of psychological equality.

SOME COMMON CHARACTERISTICS OF PSYCHOPHYSICAL METHODS

If you survey the techniques used in psychophysical experiments you will find what appears to be an extremely large number of them. Actually, however, most of these are only variations of the three different methods we shall discuss here. But first a few words about some of their common characteristics.

The observer is best used as a simple detector or null indicator. A general feature of the most common kinds of psychophysical procedures is that they require the observer to make only some very simple kinds of judgments, the simplest being: *A* is there or is not there. Another easy kind of judgment is that *A* equals *B*, or *A* does not equal *B*. Man is basically a good detector and null indicator. He can easily decide whether he has a sensation or doesn't, and he can often tell with great precision whether two sensations are equal. The psychophysical methods are designed to capitalize on these abilities.

Another slightly more difficult kind of judgment is: *A* equals *B*; *A* is greater than *B*; or *B* is greater than *A*. Telling the *direction* of a difference is a little harder than deciding only that there is a difference, although generally speaking this is something which the human can also do with good accuracy.

More complicated kinds of judgments—*A* is twice as great as *B*; *C* is half of *D*; or the difference between *E* and *F* is the same as the difference between *F* and *G*—such judgments are made only with much greater difficulty. Although these judgments can be made with fair consistency in certain specialized situations, these situations are of limited human engineering interest.

The most difficult kind of judgment is one that requires man to behave as a meter—to give a reading (or judgment) about the value of some stimulus in absolute terms. When you point a photoelectric light meter at a light source, the needle comes to rest at some value on the scale, say, 40 foot-candles. If you repeat the measurement there will be some variation in the readings you get, but the variation will usually be trivial. If you now try the same kind of measurement by asking a person to estimate the amount of illumination in foot-candles, or in any other units, the errors in his judgment will be enormous. This is why even the professional photographer depends on a photoelectric exposure meter to tell him the amount of light in a scene—he knows he cannot trust his own ability to make this kind of judgment.

Thresholds vary. Perhaps the most conspicuous characteristic of thresholds is that they vary. A threshold is not a sharply defined point which remains stable even for one person over a relatively short period of time. Although we earlier defined a threshold as a point, it might have been more appropriate to have called it a blurred, ragged edge. For this reason the second common feature of all psychophysical measurements is that the thresholds they are concerned with are statistical concepts.

The variability inherent in biological phenomena often discourages the engineer who first meets it. This, of course, is another instance of what we meant when we talked earlier about the difficulties of studying people. Psychologists, who have to live with this kind of variability, are not so dismayed by it, because the situation is not quite as bad as it seems at first glance. The reasons for this are, first, that statistical measures acquire a considerable amount of stability when they are based on a reasonable sample of data, and, second, that the variability in thresholds is often much smaller than the full range of the function in which the experimenter is interested.

Let us look at an illustration of the latter point. We spoke earlier (page 177) about the phenomenon of dark-adaptation—the progressive increase in the sensitivity of the eye when it is in the dark. The determination of a dark-adaptation curve involves the repeated measurement of the absolute threshold while the eye is adapting to the dark. In one study this was done by allowing the subject to manipulate a knob connected to a neutral wedge (a graded-density filter). Movements of the wedge either decreased or increased the intensity of the light the subject was looking at, and, in addition, a pencil attached to the wedge made a record of the movements on a moving drum. The subject's task was to keep adjusting the position of the wedge so that the light was just barely visible at all times. As we shall soon see, this is a variation of the method of adjustment.

The results of one dark-adaptation trial are shown in Figure 41. As such records go this is a rather bad one, because these are raw data for only a single run. Since no statistical averaging has been done here, we have only a single observation at any point in time. Even so, the variability is small compared to the range of the curve itself. Since this is a logarithmic scale, the range of visual sensitivity shown here is of the

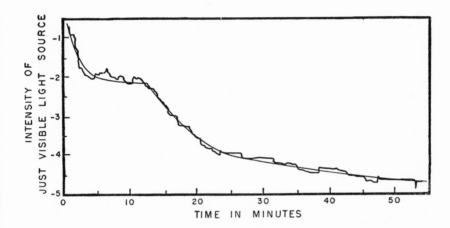

FIG. 41. The ragged line is an unsmoothed curve of dark-adaptation recorded for one subject using a psychophysical procedure modified from the method of adjustment. The smooth curve is an average line drawn through the data by visual inspection. This illustrates the normal range of variation one usually gets in experiments of this sort. (Data of Craik and Vernon[38] from Chapanis.[27])

order of 10,000 to 1. The maximum amount of deviation from a smooth curve is about 0.25 log units—or a factor of less than 2 to 1. A factor of 2 to 1 looks big in absolute terms, and it is this evidence of instability that vexes the physical scientist. Compared to a factor of 10,000 to 1, however, this instability seems rather trivial.

This is only one illustration out of many we could have picked to make the point that, in terms of its practical significance, threshold variability is not as bad as it looks at first. The other point you should keep in mind is that thresholds become a lot more stable whenever we take a number of measurements and do some averaging.

THE PSYCHOPHYSICAL METHODS

With these general considerations in mind, let us now have a look at the basic psychophysical methods.

THE METHOD OF ADJUSTMENT

The method of adjustment—method of reproduction, method of equation, or method of average error, as it is variously called in the psychological literature—is one of the oldest and most basic psychophysical methods. Although it is especially well suited to the measurement of the absolute threshold and of equalities, it can also be used for the measurement of differential thresholds. When it is used for equality measurements, the subject is given control of the stimulus dimension and is allowed to vary the stimulus until it matches some standard. If the method is used for the measurement of an absolute threshold, the subject is usually required to vary the stimulus until it is just perceptible. Repeated settings are made by the subject, and the average of these settings gives the measurement we want.

Deriving the luminosity curve of the eye illustrates a practical problem which has been attacked with the method of adjustment. Two adjacent photometric viewing fields are set up and one field, the standard, is filled with light of, let us say, 610 millimicrons. The other field, the variable, is filled with light which is slightly less reddish, let us say, 605 millimicrons. The observer now varies the intensity of the variable field until he judges that it looks just as *bright* as the standard. If the standard has a radiant emittance of one watt per square meter, ten or more equations of this sort would yield successive settings of, perhaps, 0.88, 0.92, 0.90, 0.95, 0.90, 0.92, 0.86, 0.88, 0.91, and 0.89. The average of these values, 0.90, tells us that about 10 per cent less radiant energy of the shorter wave length is required to produce the same sensation (brightness) as unit radiant energy of the longer wave length. A procedure very much like this is used by illuminating engineers in making photometric measurements with instruments like the Macbeth Illuminometer.

The dark-adaptation experiment by Craik and Vernon, to which we referred earlier, illustrates a variation of this method for the measurement of an absolute threshold. Another variation of the method of adjustment takes direct control of the stimulus away from the subject. In their study of the effect of screen brightness on the detectability of targets on a radar screen, Williams, *et al.*, had the subject communicate by telephone with the experimenter, who varied the intensity of target pips in accordance with the subject's instructions.[114]

The method of adjustment has been useful in work with targets on radar screens, because trained observers ordinarily need to make adjustments over only a narrow range of signal intensities. This minimizes the decay and build-up effects which tend to contaminate measurements made on phosphorescent screens. In addition, trained observers can get measurements in a relatively short period of time so that the drift of electrode voltages, which is often characteristic of radar circuitry, does not appreciably disturb the measurements.[113]

THE METHOD OF LIMITS

This method is also given a lot of different names in the psychological literature—method of just-noticeable differences, method of equal-appearing intervals, method of serial exploration, and method of minimal change. Whatever you call it, this method is well adapted to the determination of equalities and both absolute and differential thresholds.

Measurement of an absolute threshold. Reduced to its essentials, the experimenter starts with a stimulus value well above threshold and diminishes this value by small, but equal steps. At each value of the stimulus, the subject reports whether he can hear, see, smell, feel, or taste the stimulus (depending on what it is). When the subject can no longer sense the stimulus, the series is stopped. The next series usually starts with the stimulus well below threshold, and increases in value by small, but equal steps until the subject reports that he can detect it. Such ascending and descending series are repeated several times, and the average point of transition between response and no response gives the threshold.

The masking effect of noise on a pure tone. To illustrate this method consider an experiment on the masking effect of noise on the perception of a pure tone. You know, of course, that it is harder to hear speech and other sounds in a noisy environment than in a quiet one—the noise *masks* or conceals the sound we want to hear. One way of measuring this effect is to use a noisy background of constant intensity and to

Research Techniques in Human Engineering

measure the intensity of a sound which is just barely audible in such an environment. This involves the measurement of a *masked threshold* which resembles in some ways an absolute threshold. The background

TABLE 38. *A sample record sheet of ten trials made in the investigation of the masking effect of white noise. A + means that the subject could hear the test tone; a − that he could not*

Trial		1	2	3	4	5	6	7	8	9	10
Direction		Desc.	Asc.	Desc.	Asc.	Desc.	Asc.	Desc.	Asc.	Desc.	Asc.
Intensity level in *db*	*db* attenuation										
78	0	+								+	
76	2	+								+	
74	4	+				+				+	
72	6	+		+		+				+	
70	8	+		+		+		+		+	
68	10	+		+		+		+		+	
66	12	+		+		+		+		+	+
64	14	+	+	+	+	+		+	+	+	−
62	16	−	−	+	−	−		+	−	+	−
60	18		−	−	−		+	−	−	−	−
58	20		−		−		−		−		−
56	22		−		−		−		−		−
54	24		−				−		−		
52	26						−				
50	28						−				
48	30						−				
Threshold		63	63	61	63	63	59	61	63	61	65

D = 61.8   A = 62.6   T = 62.2

noise is white noise with an intensity of, let us say, 80 decibels. The test tone has a frequency of 1,000 cycles per second. Its intensity is controlled by a step attenuator which gives us variations of 2 decibels. A series of trials under these conditions might give results like those in Table 38. Note that the threshold for any particular series is taken as the midpoint between the last + and the first −, or vice versa. The average of all these individual thresholds gives us the value we want.

Although we shall say more about the problem later when we come to discuss sources of error in the psychophysical methods, we may note now that ascending and descending series of trials rarely give the same result. This is the reason why both kinds of trials are used in most problems. In working with radar screens, however, it may be necessary to compromise accuracy with the realities of the situation. There is troublesome (troublesome only if you want to make psychophysical measurements, that is) persistence of bright spots on most radar screens because of the characteristics of the phosphorescent material. This persistence would bias the data if descending series of trials were used. Thus, some radar studies (Bartlett and Sweet,[10] for example) have used only ascending series of signal intensities.

Measurement of a difference threshold. To measure a difference threshold the subject is told to report whether each value of the variable is more than, equal to, or less than the standard. The data would resemble those in Table 38 except that a typical trial would now contain some equal signs, for example, + + + + + + = = = = −, and − − − − = = = +. In each such trial there is a transition from + to = judgments and another transition from = to − judgments. The stimulus distance between these two transition points is the range of equality judgments and it covers a span of two *DL*'s: from the *point of subjective equality* to the higher differential threshold and from the same point to the lower differential threshold. As in Table 38 an estimate of 2*DL* is made for each trial. These are averaged across trials, and the average is divided by 2 to provide the difference threshold.

THE METHOD OF CONSTANTS

The method of constants is generally regarded as the most accurate and most widely applicable of all the psychophysical methods. It can be used for the measurement of equalities, of absolute and differential thresholds, and in the determination of many other kinds of psychological values which do not properly fall in the domain of psychophysics. Briefly, this is what the method involves:

From preliminary tests the experimenter selects a small number of stimuli, usually from four to seven, which he presents to the subject a large number of times, usually from 50 to 200 times each. The order in which the stimuli are to be presented is carefully randomized in advance and is unknown to the subject. If the method is to be used for measuring an absolute threshold, no standard is required. If a differential threshold, or equality measurement, is to be made, there is a standard included.

A bandwidth transmission problem. To illustrate the use of this method, let us turn to an example from the telephone company. Many television programs are transmitted from city to city via lines leased from the telephone company. The telephone company's aim is to provide as good cables and circuits as are necessary for the faithful transmission of TV signals. On the other hand, since it is very expensive to provide topnotch circuits, there is no need to provide better facilities than can be detected by the ultimate consumer. One of the electrical parameters which affects the quality of TV pictures is the bandwidth of the transmission channel. A reduction in bandwidth generally reduces the sharpness of the picture. Thus, a practical question is: How much can the bandwidth be reduced before there is a detectable increase in the fuzziness of the picture?

By some very rough exploratory trials, the experimenter decides that a bandwidth of 10 megacycles produces a picture which is indistinguishable from a top-quality TV picture and that a bandwidth of 4 megacycles produces a rather noticeable increase in fuzziness. The purpose of the experiment, then, is to determine precisely where the threshold for fuzziness lies between these crudely-determined limits of 4 and 10 megacycles. To run the experiments with real circuits would be extremely costly. Fortunately, it turns out that a very good simulation of the effects of bandwidth restriction can be achieved by optical defocusing. The two bandwidth values referred to above correspond to optical defocusing of 0 and 10 mils movement of the projection lens. The precise conversion of optical defocus into bandwidth need not concern us in describing the method involved.

A series of trials is now set up in which an observer views two identical scenes, one a standard picture, the other a partly blurred one. The amounts of blurring are set up in steps of 2 mils, and the task of the subject is to report which of the pair of pictures is the sharper. The order of the trials is varied so that the amounts of defocus might be in this order: 10, 8, 4, 10, 2, 6, 6, 0, 4, etc. Fifty trials are presented at each value, and a set of results might be like those in Table 39 and Figure 42.

In this experiment the subject is not allowed to say that the two pictures are equal; he is forced to guess which one is the sharper. Under these circumstances, when the subject cannot discriminate a difference he will get about 50 per cent correct—about what he would get if he were to avoid looking at the pictures and depend on the toss of a coin to tell him the answer. As the amount of defocus becomes more pronounced, however, more and more of the subject's responses are correct.

The threshold is conventionally taken as the midpoint between the guessing level (50 per cent) and complete certainty (100 per cent), or 75 per cent. There are a variety of techniques available for locating the 75 per cent point, ranging from simple graphical interpolation to

TABLE 39. *A set of psychophysical data obtained with the method of constants in the study of the just-noticeable fuzziness of TV pictures (The problem and method are genuine; the data are fictitious)*

Mils defocus of the picture (arbitrary units)	Percent of fuzzier pictures correctly identified
10	98
8	94
6	88
4	80
2	62
0	48

solutions by means of least-square curve fitting. Although the different methods give slightly different results, the variations due to methods of locating the threshold are generally small when compared with the difference between subjects, or the magnitude of other variables and parameters in the experimental situation. It is worth noting, however, that the results of such experiments frequently yield linear plots on arithmetic probability paper, or logarithmic probability paper. When this is true, the task of graphical interpolation is greatly simplified. The data in Table 39 are shown in Figure 43 to lie along a good straight line when the points are plotted on arithmetic probability paper. The threshold, by simple graphical interpolation, is found to be 3.6 mils.

In this experiment the subject was forced to guess which one of the pictures was the sharper. This is an instance of the *forced-choice* method. It is not at all necessary to run trials this way. The subject might have been allowed to say either that he thought the two pictures

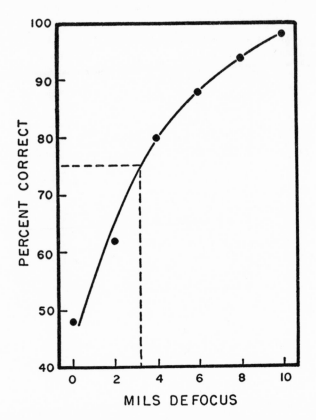

FIG. 42. Percentages of correct decisions made in judging which of two pictures is more blurred. The abscissa shows the amount of blurring introduced by optically defocusing one picture. These are the data in Table 39. (From Chapanis.[27])

were equal, or that he could see a difference in fuzziness between them. In this case, the judgments in Table 39 would be labelled "Per cent of judgments in which a difference was noted" and the values would run from about 0 per cent to about 100 per cent. The threshold would be taken as the 50 per cent point.

Context effects and the method of constants. Thresholds obtained with the method of constants are subject to a kind of distortion which may be important in some practical situations. The distortion is introduced by the context of the stimuli used to obtain the threshold. In using the method of constants the experimenter should try to pick test stimuli which are distributed symmetrically around the threshold—he tries to pick, let's say, three values of the stimulus above the threshold and

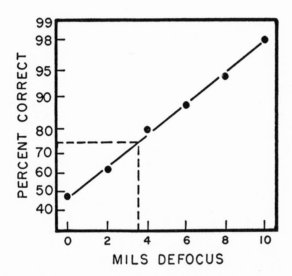

FIG. 43. These are the same data as in Figure 42 except that the percentages of correct judgments are plotted on spacings computed from a normal, or Gaussian, probability curve. (From Chapanis.[27])

an equal number below the threshold. The trouble arises because the experimenter sometimes does not know even approximately what the threshold is going to be—that is the reason he is doing the experiment. As a result, he may guess wrong and use too many stimuli above the threshold—or below the threshold—and so have a range of values which is asymmetrical around the threshold. This asymmetry has a marked effect on the number of positive and negative judgments the subject makes. Both Doughty[42] and Blackwell[13] show that such asymmetry displaces the threshold downward if there are more stimuli below the threshold; upward if there are more above the threshold.

The method of constant stimuli is also used for establishing equalities between different stimuli. For example, it is used in the investigation of illusions, that is, discovering what objective magnitudes of two somewhat different stimuli will look the same. A related kind of problem is the measurement of equal luminosity curves, or equal loudness functions. These are situations in which it may be especially difficult to establish in advance what range of stimuli will be symmetrical around the point of subjective equality.

Doughty found that the method of average error is also subject to the context effect if the subject is not allowed to vary the stimulus above and below the threshold, or point of subjective equality. In its customary form of usage, however, it is this freedom to vary the stimulus over wide limits which allows the subject to set his own context, so to speak, and so free the method of average error from contextual distortion. Although it has not been demonstrated experimentally, the method of limits should also be free of contextual distortion for essentially the same reason.

The choice of a threshold criterion. A word is in order about the use of the 75 per cent point for the determination of the threshold value. There is, of course, some arbitrariness in this value, and the experimenter may select other values if he likes. For example, in one radar study the investigators used a 90 per cent value for the threshold.[8] This is pushing the threshold up more nearly to complete certainty, and, for some human engineering purposes, this may have some advantages. However, one argument against using such high percentage values is that they yield thresholds which are, statistically speaking, less stable than thresholds computed from lower percentages. Because of the nature of most psychophysical functions, a change of a few percentage points generally produces a bigger change in the threshold at large percentage values than at those around 50 per cent. In Figure 43, for example, a change from 96 to 98 per cent changes the threshold from about 8.6 to 10 mils. A change from 73 to 75 per cent, however, only changes the threshold from about 3.3 to 3.6 mils.

Sometimes physical scientists and engineers ask why one does not take the 100 per cent point as the threshold, since this would mean that the stimulus difference can be detected all the time. There are two reasons for this, one stemming from the mathematical nature of the function, the other from the contrariness of human behavior. The first difficulty is that the point at which the psychophysical function first reaches 100 per cent is practically impossible to determine with any precision because it means the determination of a point on a curve which is becoming asymptotic to a horizontal line. The second reason has to do with the fact that people frequently do not give you 100 per cent correct responses even when you are sure that the stimulus, or stimulus difference, must be very distinct and obvious. Momentary lapses of attention, errors in writing down his own response (if the subject is doing it), your errors in writing down the subject's answers (if you are doing it), or errors in pushing the correct button (if you are using a mechanical system of recording the data), almost always seem to intrude themselves into the data. Perfect sets of psychophysical data are rare.

Forced-choice variations. As a general rule, the forced-choice method gives somewhat more precise thresholds than a method which allows the subject to say merely that he did or did not perceive a stimulus or stimulus difference.[13] One reason is that we can usually sense things much better than we think. When stimuli are near the threshold we have to make our decisions on the basis of very small, or very faint, cues. Most of us find it difficult to make decisions about such indistinct cues unless we are forced to do it. Night vision training, for example, consists mostly of teaching men to respond to minimal sense impressions. When army recruits first engage in maneuvers at night, they stumble around and have a difficult time of it. Gradually, however, they learn that one kind of blurred, indistinct blob is a man, that another kind of black shape is a tree, and that still another is a tank. At the end of a week's training they get along very well. As nearly as we can tell, the trained man does not literally have better eyes—he has merely learned to make better use of the little information available to him. This is undoubtedly what happens when choices are forced in psychophysical studies.

Forced-choices can be used only with the method of constants, but they can be used in measuring either the absolute or the difference threshold. In any case, the subject is required to guess in what temporal interval, or in what spatial position, the stimulus is presented. The latter, for example, was what the subject had to do in the study of bandwidths for TV transmission.

To take another example, let us apply the method to the study of the detectability of targets on a radar screen. We divide the face of the radar screen into eight wedge-shaped sectors and tell the subject that the signal will always appear in one of the eight sectors. The location of the sectors used in successive trials is random, and the intensity of the target pips is also randomized in the same way as the various amounts of defocus were randomized in the previous example. A tone, or other signal, tells the subject when the signal is present, and his task is to guess in which one of the eight sectors the signal appears. Notice that the subject is not allowed to say that he cannot see it; he is forced to make a guess.

In an experiment of this sort, we would expect the percentages of correct judgments to range from about 12.5 (chance level) to 100 per cent. The threshold would ordinarily be taken as that intensity which gives us about 56.25 per cent of judgments correct—half way between chance and complete certainty. Note, incidentally, that this application gets us away from some of the difficulties we mentioned earlier about doing psychophysical experiments on radar screens. By varying the location of the target from trial to trial, we are less troubled with the cumulative effects produced by repeated stimulation of the same spot on a phosphorescent surface.

Instead of varying the spatial location of the stimulus from trial to trial, we might have varied its temporal location. For example, three tones, spaced two seconds apart, could be used to mark off two time periods. The stimulus is presented either during the first or second time period, and the subject is required to guess in which one it occurred.

OTHER PSYCHOPHYSICAL METHODS

The methods we have discussed above are probably the most important ones for human engineering work. A good source of additional information on the psychophysical methods is Guilford's text.[62]

SOME SOURCES OF ERROR IN PSYCHOPHYSICAL MEASUREMENTS

Having had a brief look at the principal psychophysical methods, let us now turn to some of the common sources of error that come into measurements of this sort and, more especially, some of the precautions one should observe in planning psychophysical experiments.

Control cue errors. One of the commonest errors which biases psycho-

physical experiments done by novices arises from cues which the subject can get about the stimulus from the control. Perhaps the simplest
way of illustrating the point is to take an example. In the method of
adjustment, the subject is frequently allowed to control the stimulus—
let's say, the intensity of a light. Let's suppose that the control he has
is a knob on a slide wire rheostat. This kind of an experimental apparatus is likely to lead to the situation in which the subject, after the first
trial, sets the knob on successive trials to about the same position on
the rheostat. Oftentimes the subject does this without realizing what
he is doing. In the case of rotary rheostats—such as variacs—the subject may tend to turn the knob a constant angular amount from the
starting point. If the knob has some sort of a pointer on it, he may tend
to set the pointer to some particular direction. All of these situations
are related in that they usually produce systematic biases and reduce
the variability in the data.

A good practical illustration of the way control cue errors can affect
psychophysical measurements occurred during an experiment run by a
graduate student at Johns Hopkins. This experiment was designed to
measure the just-noticeable difference for visual flicker. In one part of
the experiment, the subject saw a standard light flickering at 10 cycles
per second (that is, it had a period of 100 milliseconds). He was required to adjust the rate of flicker of a variable light until the two appeared to be flickering at the same rate. The subject had two controls:
a double-throw switch enabled him to view either the standard or
variable flicker at will; a rotary knob enabled him to increase or
decrease the flicker rate for the variable light.

During one series of 40 trials early in the experiment, the data came
out as shown in Figure 44. The measurements are the differences between the period of the standard flicker rate and the period of the
setting for the variable light made by the subject. Notice that for the
first 26 trials there is a considerable amount of variability in the
settings, as one might expect, but that after trial 27 the errors suddenly
become much more stable, with the one exception at trial 32. The
explanation for this change in accuracy is to be found in the control
knob (Figure 45) and in the way the subject made his settings. Careful
probing by the experimenter disclosed that at trial 27 the subject
started making his settings in a systematic way. First he made a
coarse adjustment without paying attention to the knob. Having
made such a coarse adjustment, he then made a fine adjustment by
bringing one of the bumps on the knob uppermost. Even the discrepant

result on trial 32 can be explained by this system. On that particular trial, the subject brought the wrong bump uppermost. Fortunately, this flaw in technique was discovered early in the experiment. The experimenter immediately substituted a plain circular knob for the one shown in Figure 45.

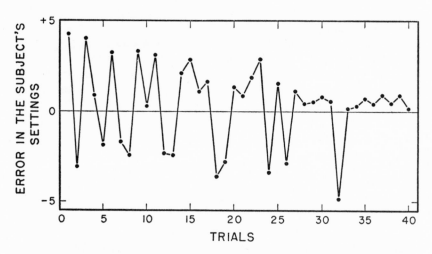

FIG. 44. Errors made by one subject on 40 consecutive trials in a psychophysical experiment designed to measure the just-noticeable difference in visual flicker rate. The method of adjustment was used. An error is the difference between the subject's setting and the standard flicker rate. Measurements are in terms of the period, in milliseconds, of the flash rate. Notice the sudden change in the consistency of the errors after trial 27. At this point the subject started making use of a control cue.

The principal ways of guarding against errors of this sort are:

1. Use a control which has no markings or other identifying characteristics which would introduce consistency in the settings. This means that most controls which vary in their linear position in space, such as levers, are not good for this purpose. Unadorned rotary knobs are good. Another good arrangement, but a more elaborate one, is to have the stimulus remotely controlled by a reversing motor. The subject is given a double-throw, center-off, non-locking switch to control the motor. So long as he holds the switch up, the stimulus increases in intensity. To decrease intensity he has to hold the switch down. If this kind of mechanism is used with the precaution noted in paragraph 3

below, the subject can get no cues about the stimulus from the stimulus control.

2. Use a control which has no identifiable high and low limiting stops. If the control is a rotary knob it should be on a friction-coupled shaft which eliminates any positive stops—like the tuning dial on most

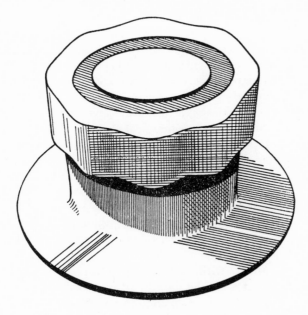

FIG. 45. The control which was responsible for the data in Figure 44.

radios. The purpose of this arrangement is to get rid of cues the subject might get from the top and bottom of the stimulus range.

3. Successive trials in the psychophysical experiment should start at different positions on the stimulus scale, and the starting positions should be set by the experimenter. In Table 38, for example, note that some of the trials started up high, some close to the threshold and some in between. Varying the starting position in this way means that the subject cannot depend on time, number of steps, number of turns of a knob, or related cues which would introduce uniformity in the settings or judgments on successive trials. When the control is in the subject's hands, variations in starting intensity can be made by putting an

additional control in series with the subject's. For example, in a study by Egan, *et al.*,[46] the subject controlled the intensity of auditory communications by means of an attenuator calibrated in 2-db steps. The experimenter had an additional attenuator in tandem with the subject's. At the beginning of each trial the experimenter set in an arbitrary value so that the subject's attenuator, in effect, had no constant base or reference value.

Space errors. When a variable stimulus is being compared with a standard, they are often placed in some sort of geometrical arrangement with respect to each other. The results one gets frequently differ depending on whether the variable stimulus is to the right, left, above, or below the standard. This is the space error.

Figure 46 shows a set of data obtained in an experiment to explore the illusion of the vertical. Vertical lines look longer than horizontal ones, and this illusion has some practical applications in certain human engineering situations. To quantify the illusion, a 6-inch standard line, oriented horizontally, was shown near the center of a large white screen. Close to the standard was a variable line which was presented in various orientations: 0° is horizontal; 90° is vertical. The psychophysical technique was essentially that of the method of adjustment.

The data in Figure 46 show a maximum illusory effect of somewhat over 10 per cent. But this is not what concerns us at the moment. Some of the trials were presented with the variable on the right of the standard; an equal number with the variable on the left. With but one exception (that is, at 90°—the data at 0° represent control data), the former arrangement shows a smaller illusory effect than the latter. In this experiment some of the trials were also presented with the variable above the standard; others with the variable below. The former arrangement produced a smaller effect than the latter, and both of these produced much smaller effects than the arrangements in which the lines were side by side. Since there are a large number of observations represented here, there is no question about the statistical significance of the effects.

There seem to be no very good explanations for most space errors; at the moment about all we can say is that this represents one of the vagaries of human behavior. Unfortunately, we cannot even say for sure which side will show the larger effects in various kinds of experiments. However, if you want the greatest generality from the results of psychophysical experiments, you will do well to vary the geometrical relationships between the variable and standard stimuli.

Time errors. In some psychophysical experiments—auditory studies being a good illustration—it is not feasible to present the standard and variable stimuli at the same time. For such studies, one tone is usually presented before the other, and the judgments are made about stimuli

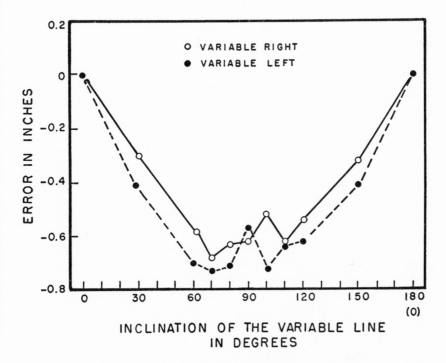

FIG. 46. Constant errors made in setting the length of a variable line so that it looked as long as a fixed horizontal line six inches long. The variable line was inclined at the various angles shown along the baseline. In half of the trials the variable line was on the right of the fixed one; in the other half it was on the left. Each point is an average of 20 settings. (Data of Pollock and Chapanis[91] from Chapanis.[27])

separated in time, rather than in space. Experiments on the judgment of weights usually present stimuli separated in time and visual experiments may sometimes use this procedure for any of a number of technical reasons. The outcomes of such procedures, however, are much more uniform than in the case of space errors: the second of two equal stimuli is almost always judged louder, heavier, or brighter, than the first. There are a number of theories to account for this phenomenon, but the facts are more important than the theories for our purposes. To get unbiased data when stimuli are presented successively, it is a good idea to present the variable before the standard half the time; the standard before the variable the other half. The orders would, of course, be randomized.

Direction errors. In Table 38 you will note that half the trials were presented in an ascending order; half in a descending order. The reason for this is that the two kinds of series do not generally give the same results. When the subject has been giving a series of "yeses" or "noes" he tends to persist in the same kind of response. Thus, descending trials frequently give lower thresholds than ascending ones.

Sequential response tendencies. Another source of error which comes into psychophysical experiments—especially those using the method of constants—arises from the fact that people come with certain peculiar response habits built in. If you pick a stimulus intensity which is just about at a subject's absolute threshold and present the same stimulus repeatedly, you will find correlations between successive responses. These have been extensively studied recently, and we know some of the things that people do and do not tend to do. For example, people tend to repeat responses they have just made.[98, 100, 109, 110] Still other response habits show themselves when the intensities of the stimuli are varied. Incidentally, the person seldom realizes that he has these responses tendencies. He may, in fact, believe that he is responding in a thoroughly random fashion.

The seriousness of this source of error is not quite as great as may seem at first hand. In the typical psychophysical experiment, the subject's responses are largely determined by the stimulus; it is determined much less by the preceding response. Nonetheless, it is something to keep in mind and guard against. Perhaps the best way of avoiding bias due to this source is to insure that the sequences of stimulus intensities are determined by a table of random numbers. This will mean that the subject is not required to make threshold or near-threshold responses many times in succession. In addition, response tendencies, if there are any, will be evened out throughout a whole string of data and will not introduce bias in one way or another. Incidentally, do not try to make up a random sequence out of your head—a recent experiment has shown quite conclusively that people cannot be random if they are told to be, even if they have studied statistics and probability and are familiar with tables of random numbers.[26] You will find tables of random numbers in books by Fisher and Yates[48] and the Rand Corporation.[93]

Bias due to expectancies. Making threshold judgments is difficult because, by definition, they are judgments about stimulus differences which the observer can hardly notice. As a result, observers try to find some basis on which to make consistent decisions. If the experimenter, by a word, nod, or other action, indicates what he thinks the observer *should* be doing, the observer, consciously or unconsciously, will try to comply. If the observer builds up some expectancies about the task, these will influence his judgments.

Senders and Sowards demonstrate how instructions can modify the observer's expectancies in a psychophysical experiment and alter his judgments as well.[100] They presented subjects with a light flash and a tone, both of which appeared simultaneously. The observers, however, did not know this. They were told that they were to guess which one of the two stimuli appeared first. In some trials, the observers were not told anything more than this. In others, the experimenter said, "To help you, I will tell you that in this experimental session the ratio of simultaneous to non-simultaneous presentations is one-to-one (or two-to-one, or four-to-one). That is, for every trial in which the onset of the light and tone are simultaneous, there will be one trial (or two, or four trials) in which they are not." Notice that the subject was not told that he should make his responses in these ratios; he was merely told that the stimuli would be in these ratios. Actually, you remember that the two stimuli were always presented simultaneously.

The results of this experiment show dramatically how the subject's responses were affected by these instructions. Subjects tended to give responses agreeing with the ratios stated by the experimenter.

To sum up, then, the experimenter must take special pains to avoid biasing the subject in psychophysical experiments. Observers in psychophysical experiments are always asking questions: "Was that one brighter than the one before?"; "Was that a blank one?"; or "Are you giving me a whole string of hard ones?" They are looking for some kind of clue—any kind of clue—about how well they are doing. Unfortunately, it is very difficult to reply to such questions in a completely non-committal manner. But unless he does so, the experimenter runs the risk of influencing the observer's judgments.

Chapter 8

Modern civilization depends for its very existence on the rapid transfer of information from one place to another. Prominent among the techniques for doing this is the speech communication system. Reduced to its bare essentials such a system involves a talker, some form of electronic system, and a listener at the other end. The information to be transmitted is encoded in the form of audible speech. Examples of speech communication systems are: (1) Radios—ground-to-ground, ground-to-air, air-to-ground, and air-to-air; (2) telephones —residential, ship-to-shore, railroad-to-railroad, or other type; and (3) annunciator systems—between one room and another, or between one building and another. All of these systems create many human engineering problems. Noise, static, the fidelity with which speech sounds are transformed into electrical energy, the bandwidth allocated to the communication channel—all these and many more factors contribute to the efficiency of the system.

In designing speech communication systems the engineer frequently wants to know, "How good is this system?" or "Is circuit A better than B?" The criterion used most frequently for evaluating the effectiveness of a speech communication system is the percentage of speech sounds (syllables, words, or sentences) which can be correctly conveyed over it. This is typically measured by an *articulation test*, and the result of such a test is called an *articulation score*. The terms *intelligibility testing* or *tests of speech intelligibility* are often used synonymously with articulation testing.

Because the articulation test involves people (talkers and listeners) and equipment (microphones, amplifiers, earphones, and so on) it is a human engineering technique in the full sense of the word. This chapter is devoted entirely to the problems of articulation testing.

PURPOSES OF ARTICULATION TESTING

Articulation tests were devised originally by telephone engineers for measuring the effectiveness of telephonic communication. During World War II these tests were refined and applied to many other systems. Today they are used widely for the four major purposes described below.

COMPARISON OF COMMUNICATION SYSTEMS AND COMPONENTS

The most important use of articulation tests is in solving problems of equipment design. For example, these tests are used to evaluate the comparative effectiveness of different microphones, loudspeakers, amplifiers, and entire systems. Although such equipments are usually designed to have certain assumed theoretical advantages, the proof of the pudding is in the eating and the engineer may want to know: Is there any practical difference between two or more equipments in terms of the speech intelligibility which can be obtained with them?

In addition, the designer may sometimes want to investigate the effects of certain experimental variables on speech communication. For example, frequency and amplitude distortion introduced by transducers and amplifiers frequently can improve speech intelligibility under certain conditions. Techniques for shielding microphones and earphones from noise in the environment are also problems of some interest to the designer.

DETRIMENTAL EFFECTS OF ENVIRONMENTAL FACTORS

Modern civilization is a noisy one and man-made interferences constitute serious hazards to effective communication. In addition to man-made interference there is a variety of noises and interferences created by the environment or the device itself. Static and internal noise originating from amplifiers and other components are of this nature. The detrimental effects of these noises and the amounts which can be tolerated in communication circuits are usually measured by articulation tests.

Occasionally the problem may be reversed: The designer may be interested not in improving speech intelligibility but in destroying it. This is the primary purpose of "jamming." The engineer attempts to produce devices which will effectively distort, conceal, or mask communications which other people are trying to engage in. The articulation test is again the criterion most frequently used for evaluating how well various types of interference work.

Communication at high altitudes where the air is thin, or through water, are other special environmental problems of great interest to the human engineer.

SELECTING SPECIAL VOCABULARIES

Some words are more intelligible than others simply because of the kinds of sounds that are contained in them. For this reason it is often effective in human engineering work to devise special vocabularies for communication purposes. An example is the ICAO (International Civil Aviation Organization) Word Spelling Alphabet. This vocabulary, devised after years of intensive research, has the property of being intelligible even when spoken by persons of different nationalities and with diverse dialects. The articulation method is again the primary technique used for constructing such vocabularies.

RATING AND TRAINING OF PERSONNEL

There are large differences between people in their inherent talking and listening abilities. Some people are very clear talkers, while others mumble their words or otherwise communicate ineffectively. The same is true of listeners. Some are extremely good; others are not so adept at receiving spoken communications. Articulation tests provide a convenient way of testing people to discover their inherent abilities in this respect and for giving them training in the use of communication equipment.

FACTORS IN THE DESIGN OF
ARTICULATION TESTS

From the discussion above you can see that properly designed articulation tests are a powerful help in solving many practical design problems involving communication systems. In order for these tests to be valid, to insure comparability of results from one equipment to another, and to avoid sources of bias and error, certain important factors must be considered in their design. These are:

1. The selection of personnel to conduct the test.
2. Training of the personnel who participate in the test.
3. The choice of sounds used for the test.
4. The procedures used in administering the tests.
5. The choice of test equipment and test conditions.
6. The experimental design used for the tests.

PERSONNEL

One of the most critical items necessary for the conduct of valid and meaningful articulation tests is the selection of proper personnel to participate in the tests. Two kinds are needed: talkers and listeners. Some general requirements which both groups must satisfy are:

Naïvete. Both talkers and listeners should be naïve with regard to the communication systems under test. The reason for this should be obvious. If talkers know what kinds of systems are under test, and if they have personal preferences for one or another type of system, they might very easily bias the outcome of the experiment—by the way they talk, by the intensity of the spoken words, by the inflection given certain words, and so on. Similar considerations apply to the listeners. Above all, never use communication engineers, or anyone who has had anything to do with the design of any of the systems.

Education. The educational level of both talkers and listeners should be high enough so that both are familiar with the words to be used in the testing program. Samples of test words are given later in this chapter.

Intelligence. Talkers and listeners should be intelligent enough to understand the general purpose of the tests. They should also be able to understand the proper use of the various equipments—earphones, microphones, loudspeakers—in order to give them a fair trial.

Training. Both talkers and listeners should receive training in the proper use of the equipment. This is particularly important for certain types of communication systems. For example, if the tests involve such items as noise-cancelling microphones or microphones with noise shields, it is especially critical to the tests that the microphones be used properly. Earphones, especially if they are encased in ear muffs or noise-protective cushions, should be correctly fitted to the listeners and the listeners should be trained in the proper use of these items.

If words are to be used as test materials, both talkers and listeners should be familiar with the words which will appear during the tests. This does not, however, apply to test materials consisting of sentences or questions. See also the section on training starting on page 283.

Use. Talkers should not be used as listeners, nor should listeners be used as talkers. People should be selected for a particular job and used only at this job.

Motivation. Both talkers and listeners should be highly motivated to do as good a job as possible. Full-scale articulation tests are tedious, time consuming, and often monotonous. Inattention, boredom, and fatigue generally tend to produce highly variable data which make it harder to evaluate the performance of the systems under test. Highly motivated subjects tend to give more stable data than do poorly motivated ones.

Some additional requirements for talkers. In addition to the general requirements listed above, the talkers should have no obvious speech defects nor should they have any noticeable regional or national accents. As is typical with most such general rules there are occasions when these requirements may be deliberately violated. Earlier in this chapter, I mentioned the ICAO Word Spelling Alphabet. This was devised after extensive tests in which talkers with marked national accents read words over various communication systems. The purpose of the tests was to find words which would be comprehensible even when spoken by persons with widely different accents. In this case, talkers were deliberately selected to satisfy this requirement.

Some additional requirements for listeners. The listeners should have reasonably normal hearing; that is, they should have no hearing losses greater than 15 or 20 db over bands more than an octave wide in the important speech range (roughly from 200 to 3,000 cps).

On the number of subjects. It is difficult to specify exactly the number of people who should be used in articulation tests. As a general rule the greater the number of talkers and listeners the more valid and general the findings. On the other hand, a large number of subjects greatly lengthens the tests and makes them much more expensive to run. The compromise at which the practical experimenter aims is to use only as many talkers and listeners as are necessary to get meaningful results.

Some other general rules which are relevant to this problem follow:

1. Rough estimates of large differences can be obtained with relatively few subjects. Getting precise results on small differences takes many more subjects.

2. Fewer talkers than listeners are needed for most kinds of tests. An exception is in the testing of microphones. For example, talkers with fat, heavily padded necks might have great difficulty getting words through to a throat microphone. As another example, certain facial contours (particularly of the nose and chin) produce difficulties with close-talking microphones. In both of these cases the experimenter should use a representative sample of talkers to get good *average* data.

3. Increasing the number of listeners increases the task of scoring the tests, but does not materially increase the time and labor of running the tests. The reader can verify this from the experimental design in Table 43.

4. As a very rough guide, the general experience of people who conduct articulation tests is that a minimum of four talkers and six listeners should be used in most instances.

TRAINING

As with most other things in life, the ability to talk and listen improves with practice. For this reason a considerable practice period should precede the formal testing of any new equipment.

Training in the proper use of equipment. One of the principal reasons for practice is to acquaint both the talkers and the listeners with the proper use of the equipment—microphones, headsets, ear muffs, helmets, and other items. For example, the way in which test personnel use microphones has an important influence on the outcome of articulation tests. In order to control the speech input to the microphone throughout all tests it is important, among other things, to keep the base of the microphone at a constant distance from the lips of the announcer (See Figure 47). Similar precautions apply to many other pieces of equipment.

Training in talking level. The talker's voice level is another important factor affecting the outcome of articulation tests. One purpose of the

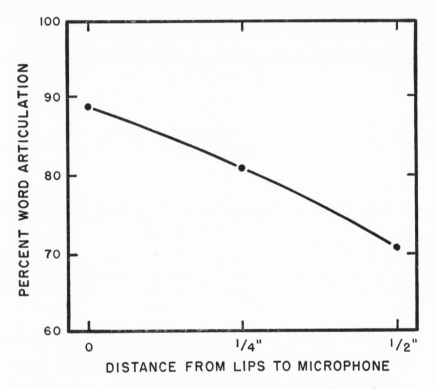

FIG. 47. Word articulation scores as a function of the distance between the microphone and the talker's lips. These tests were made with talkers and listeners in 115 db of simulated aircraft noise.[86]

practice sessions should be to train each talker in keeping the intensity level of his voice as constant as possible.

The amount of training. In general, training should continue with a communication system until it has become entirely routine. That is, both talkers and listeners should be completely familiar with the testing situation and with the exact procedure used in the test. One way of finding out when this has been achieved is to plot the mean articulation score of the crew over a period of time. When the mean score shows no further improvement over a series of several tests, you may assume that the crew is ready to go. Figure 48, for example, shows a typical learning curve obtained by a listening crew at the Harvard Psycho-Acoustic Laboratory. This crew showed a substantial improvement

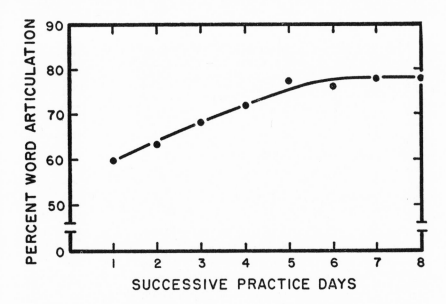

FIG. 48. Learning curve for a crew of 10 listeners. Each point is an average of 12 tests. Tests were read over an interphone system by three well-trained talkers. Talkers and listeners were in an intense noise (115 db) simulating that of a propeller-driven airplane.[86]

for the first five days of practice and then settled down to a relatively stable level of performance thereafter. These practice sessions were two to three hours long each day.

When two or more different equipments or systems are being tested the training period should be continued until a stable *difference* (or differences) has been established between the average scores obtained with the various systems. The crew is ready for a series of formal tests only after this stable level of performance has been reached.

Although a crew may have been thoroughly trained for one set of conditions it may require some additional training if tests are to be made with radically different equipment, environmental noise, or test procedures. Once a crew has been thoroughly trained, however, any such additional practice is usually of short duration.

TEST MATERIALS

As a result of practical experience, engineers and scientists have developed four main classes of test materials: (a) nonsense syllables; (b) monosyllabic words; (c) spondaic words; and (d) sentences. Each of these has its own characteristics which make it suited for certain kinds of work.

Nonsense syllables. Nonsense syllables are much superior to words or sentences when you want to determine accurately the effectiveness of a device in transmitting particular speech sounds. Since nonsense syllables are constructed by taking the fundamental speech sounds in various combinations, the relative frequency of each of these speech sounds can be adjusted to suit the requirements of the test.

The primary disadvantage of nonsense syllables is that the testing crew must be thoroughly trained in their use. The talkers must pronounce the speech sounds correctly and the listeners must have extensive practice in recording (in phonetic symbols) the sounds they actually hear. For most practical purposes the engineer or designer will want to use one of the other kinds of test material discussed below. If you should be interested in the use of nonsense syllables for special test conditions you will find a more thorough discussion of them in Beranek's book.[11] Beranek also gives 4 lists of 84 nonsense syllables; a Psycho-Acoustic Laboratory report gives 8 such lists.[4]

Monosyllabic words. Monosyllabic words are the test materials most commonly used in articulation tests. Years of research and testing have led to the development of specialized lists which satisfy a number of stringent requirements:

1. Each list covers a wide range of difficulty so that it can be used for most types of articulation comparisons. That is, within each list there are some easy, some moderately difficult, and some very difficult sounds represented.

2. The *spread* of difficulty is about the same in each list, and each list has very nearly the same *average* difficulty.

3. Each of the lists has words with phonetic composition very similar to that of the English language.

4. Rare and unfamiliar words have been avoided as much as possible.

Because of the care which has gone into their compilation (especially as regards the speech sounds represented) these lists are called phoneti-

TABLE 40. *Two lists of phonetically balanced monosyllabic words for use in articulation tests*

PB-50 List 1		PB-50 List 2	
1. are	26. hunt	1. awe	26. nab
2. bad	27. is	2. bait	27. need
3. bar	28. mange	3. bean	28. niece
4. bask	29. no	4. blush	29. nut
5. box	30. nook	5. bought	30. our
6. cane	31. not	6. bounce	31. perk
7. cleanse	32. pan	7. bud	32. pick
8. clove	33. pants	8. charge	33. pit
9. crash	34. pest	9. cloud	34. quart
10. creed	35. pile	10. corpse	35. rap
11. death	36. plush	11. dab	36. rib
12. deed	37. rag	12. earl	37. scythe
13. dike	38. rat	13. else	38. shoe
14. dish	39. ride	14. fate	39. sludge
15. end	40. rise	15. five	40. snuff
16. feast	41. rub	16. frog	41. start
17. fern	42. slip	17. gill	42. suck
18. folk	43. smile	18. gloss	43. tan
19. ford	44. strife	19. hire	44. tang
20. fraud	45. such	20. hit	45. them
21. fuss	46. then	21. hock	46. trash
22. grove	47. there	22. job	47. vamp
23. heap	48. toe	23. log	48. vast
24. hid	49. use (yews)	24. moose	49. ways
25. hive	50. wheat	25. mute	50. wish

cally-balanced, or, in short, PB lists. Two lists of fifty words each are given in Table 40. Fifty is the smallest number of words which will satisfy the requirements listed above. Beranek gives 8 lists of 50 words each and the Psycho-Acoustic Laboratory report gives 20 such lists.

Spondaic words. Spondaic words, or spondees, are words which have two syllables spoken with equal stress on both syllables. Spondees are particularly useful when the purpose of the articulation test is primarily to establish accurately the amplification or power level at which speech can just be heard. The spondaic words in each list in Table 41 are very nearly of exactly the same difficulty. As a result these words all reach the threshold of hearing within a very narrow range of intensity. This enables the experimenter to measure the threshold of hearing with great precision.

TABLE 41. *Two lists of spondaic words for use in articulation tests*

Spondee List 1		Spondee List 2	
1. airplane	22. hardware	1. although	22. mushroom
2. armchair	23. headlight	2. beehive	23. nutmeg
3. backbone	24. hedgehog	3. blackout	24. outside
4. bagpipe	25. hothouse	4. cookbook	25. padlock
5. baseball	26. inkwell	5. cookbook	26. pancake
6. birthday	27. mousetrap	6. daybreak	27. pinball
7. blackboard	28. northwest	7. doormat	28. platform
8. bloodhound	29. oatmeal	8. duckpond	29. playmate
9. bobwhite	30. outlaw	9. eardrum	30. scarecrow
10. bonbon	31. playground	10. farewell	31. schoolboy
11. buckwheat	32. railroad	11. footstool	32. soybean
12. coughdrop	33. shipwreck	12. grandson	33. starlight
13. cowboy	34. shotgun	13. greyhound	34. sundown
14. cupcake	35. sidewalk	14. horseshoe	35. therefore
15. doorstep	36. stairway	15. hotdog	36. toothbrush
16. dovetail	37. sunset	16. housework	37. vampire
17. drawbridge	38. watchword	17. iceberg	38. washboard
18. earthquake	39. whitewash	18. jackknife	39. whizzbang
19. eggplant	40. wigwam	19. lifeboat	40. woodchuck
20. eyebrow	41. wildcat	20. midway	41. workshop
21. firefly	42. woodwork	21. mishap	42. yardstick

Sentences. Sentences have only limited usefulness in testing communication equipment. There are several reasons for this. First, the intelligibility of sentences is determined to a large extent by such factors as meaning, context, rhythm, and interest. As a result, articulation scores obtained with sentences are generally so high that communication systems must differ greatly before any substantial difference shows up in the scores. In addition, the influence of psychological factors on articulation scores obtained with sentences makes the results difficult to analyze and interpret. Finally, the listeners soon learn the sentences. For this reason the experimenter needs a very large number of different sentences if he plans an extensive testing program.

Despite these disadvantages, sentences are useful in certain special circumstances. For example, if you are testing the intelligibility of telephone talkers, sentences require the talker to engage in a more complex type of action than would the speaking of single words. Rate of talking, inflection, stress pattern, and maintenance of a consistent loudness level can be tested adequately only with sentence material.

Two main forms of sentences are used in articulation tests. In one form of test the listener is required to respond to simple questions or commands by an appropriate word or phrase. Each sentence is then scored right or wrong depending on whether the listener understood the meaning of the question.

In the second type of test the listener is required to write down the sentence as read to him. The articulation score is then based on the number of key words correctly recorded by the listener. This method provides a more objective measure of the words actually heard. In the lists in Table 42, the five key words to be scored in each sentence are italicized. The scientists who constructed these sentences made special efforts to avoid clichés, proverbs, and other stereotyped sentences, as

TABLE 42. *Two lists of sentences for use in articulation tests. Only the italicized words are scored*

List 1

1. The *birch canoe slid* on the *smooth planks*.
2. *Glue* the *sheet* to the *dark blue background*.
3. *It's easy* to *tell* the *depth* of a *well*.
4. These *days* a *chicken leg* is a *rare dish*.
5. *Rice* is *often served* in *round bowls*.
6. *John* is *just* a *dope* of *long standing*.
7. The *juice* of *lemons makes fine punch*.
8. The *chest* was *thrown beside* the *parked truck*.
9. The *hogs* were *fed chopped corn* and *garbage*.
10. A *cry* in the *night chills my marrow*.
11. *Blow high* or *low* but *follow* the *notes*.
12. *Four hours* of *steady work faced* us.
13. A *large size* in *stockings* is *hard* to *sell*.
14. *Many* are *taught* to *breathe through* the *nose*.
15. *Ten days' leave* is *coming up*.
16. The *Frenchman* was *shot when* the *sun rose*.
17. A *rod* is *used* to *catch pink salmon*.
18. He *smoked* a *pipe until* it *burned* his *tongue*.
19. The *light flashed* the *message* to the *eyes* of the *watchers*.
20. The *source* of the *huge river* is the *clear spring*.

TABLE 42 (CONT.)

List 2

1. *Death marks* the *end* of *our efforts.*
2. The *gift* of *speech* was *denied* the *poor child.*
3. *Never kill* a *snake* with your *bare hands.*
4. *Kick* the *ball straight* and *follow through.*
5. *Help* the *woman get back* to her *feet.*
6. *Put* a *dot* on the *i* and *sharpen* the *point.*
7. The *hum* of *bees made Jim sleepy.*
8. A *pint* of *tea helps* to *pass* the *evening.*
9. *Smoky fires lack flame* and *heat.*
10. The *soft cushion broke* the *man's fall.*
11. *While* he *spoke,* the *others* took their *leave.*
12. The *core* of the *apple housed* a *green worm.*
13. The *salt breeze came across* from the *sea.*
14. The *girl* at the *booth sold fifty bonds.*
15. The *purple pup gnawed* a *hole* in the *sock.*
16. The *fish twisted* and *turned* on the *bent hook.*
17. A *lot* of *fat slows* a *mile racer.*
18. *Press* the *pants* and *sew* a *button* on the *vest.*
19. The *swan dive* was *far short* of *perfect.*
20. *James tried* his *best* to *gain ground.*

---

well as the very frequent use of particular words. Beranek gives 8 lists of 25 sentences each;[11] the PAL report 68 lists.[4]

PROCEDURES FOR ADMINISTERING THE TEST MATERIALS

The procedures for carrying out articulation tests depend to some extent on the kind of test materials selected.

The carrier phrase. The usual procedure for administering nonsense syllables, monosyllabic words or spondaic words is to place them at the end of a carrier phrase. Some acceptable carrier phrases are:

"You will write_____."

"Write down_____."

"Please write_____."

In each case the test word follows the carrier phrase and the listeners are instructed to record only the last word in the sentence.

The carrier phrase serves three primary purposes:

1. It prepares the listener for the word and alerts him to the response he will have to make.

2. It reduces the variability in the response of certain types of equipment. For example, the carrier phrase agitates the carbon particles in a carbon-button microphone, reducing the variability in the response of the microphone by the time the test word comes along.

3. The carrier phrase provides some preliminary words to help the talker modulate his voice. This is one way of insuring that the test words are spoken at more nearly the same intensity.

Listeners should try every item. Listeners should be instructed to respond to every item spoken by the announcer. Numerous experiments show that listeners do considerably better than chance even when they think they are guessing. If subjects are not forced to guess, they frequently develop a tendency to give up too easily as the tests progress.

Frequency of testing. The test materials should not be presented too rapidly because the listeners must have ample time to write down their responses. With well-trained test crews, it is possible to use about one item every three or four seconds. Sentences must be presented much more slowly.

Familiarity with the test materials. If the test materials are monosyllabic or spondaic words—but not sentences—both the talkers and listeners should be familiar with them. Prior familiarity with the words reduces the variability obtained in typical test runs. Sentence lists should be used only once, but word lists can be used over again.

Randomizing the word lists. If word lists are used, the experimenter should randomize the order of the words within each list. No list should be given twice in the same order nor should words be given in alphabetical order. If lists were to be repeated with the same word order, the listeners would soon learn to anticipate the words in the list simply by virtue of the order of their appearance. An additional reason for randomizing the order of the words is to counterbalance for the effects of boredom, monotony, and fatigue. For example, words at the end of the list might typically be heard less well than those at the beginning simply because of the fatigue occurring during the testing session. If the sequence of words is always the same, the communication systems may not be given a fair trial with certain words and word sounds. Randomization of the words insures that in the long run each communication system is tested equally well with all of the word sounds and under approximately equal conditions of motivation.

Scoring the tests. Two important rules should be observed in scoring the listening tests:

1. Neither the listeners nor the talkers should score the finished tests. One reason for this is that it is very difficult for listeners to be objective in scoring their own test data. For example, when people score their own responses there is a great temptation for them to be lenient with ambiguous answers on the grounds that "I really meant that after all."

2. The instructions to the scorers should stress that the words are to be scored for *sound,* not for *spelling.* For example, "knot" and "not," or "know" and "no," should be scored as equivalent. Also, "folk" and a misspelled word like "foke" would both be scored equivalent because the sounds are alike. On the other hand, "rise" is not the correct answer for "rice" because these two words sound different.

CHOICE OF TEST EQUIPMENT AND CONDITIONS

The designer usually starts with a definite idea of the kinds of equipment he wants to test by means of articulation test procedures. There are, however, some general rules and precautions which should be observed in this connection.

The use of a reference system. One of the most important things to understand about an articulation score is that it is not an absolute number. Rather, it is the result of a large number of factors, each of which influences the articulation score one way or another. Articulation scores are *relative* scores. One can place great confidence in the comparative results obtained in articulation tests when two or more systems are compared under highly standardized conditions and with the same testing crews. But the results of one series of articulation tests cannot usually be compared with the results of another series of tests run at some other time, with other testing crews, with other equipment, or in another laboratory. This characteristic of articulation test results means that an equipment or communication system should not be tested by itself. It should always be tested with some reference system in order that the designer may come out with a comparison. The reference system may be a particular system in use at the moment or a high-fidelity system especially selected to yield topnotch performance.

Components must be considered in relation to other parts of the communication system. Even if he is only comparing two components, the engineer needs to consider the other parts of the communication system used in the tests. Suppose, for example, that two microphones are being compared. Microphone *A* does not transduce speech frequencies above 2,500 cycles per second, and the other, *B*, transduces all of the important frequencies of speech. If the earphones which are used in the communication system do not transduce speech frequencies above 2,500 cycles per second the test results would undoubtedly show little or no difference between *A* and *B*. On the other hand, if earphones having a wide frequency response were used for this comparison, a higher articulation score would undoubtedly be obtained with microphone *B*. As a general rule, you cannot evaluate components meaningfully unless you know the other items of equipment with which they were tested.

One of the best arrangements of equipment is a high-fidelity sy.tem with sufficient flexibility so that the experimenter can substitute test equipments (microphones, amplifiers, earphones) for corresponding components in the basic installation.

Choice of ambient environmental conditions. Most communication systems are used in the presence of noise of one sort or another. For example, noise is often picked up by the microphone. The amount and type of ambient noise which enters the system by this means depends on three main factors: the intensity and spectrum of the ambient noise; the amount of acoustic shielding provided by the design and mounting of the microphone; and the response characteristics of the microphone itself. Noise may also influence communication because of leaks in and around the earphone. The efficiency of the acoustic seal provided at the ear by different types of earphones varies over a wide range. Finally, line noise or static may enter the system through electrical interference.

If the communication system under test is to be used in a noisy environment the articulation tests should be conducted under similar conditions. In particular, it is important that the intensity and spectrum of the ambient noise to be expected in the real environment should be duplicated in the test conditions.

In general there are four main types of noise and quiet conditions under which articulation tests can be made:

1. Quiet-to-quiet: both the talker and listener are in quiet.
2. Quiet-to-noise: the talker is in quiet and the listener is in noise.
3. Noise-to-quiet: the talker is in noise and the listener is in quiet.
4. Noise-to-noise: both the talker and the listener are in noise.

Each of these four permutations can be found in one or another type

of actual field operation. To do all four types of tests you obviously
need an acoustically treated room and a noise generator.

Monitoring the talker's voice level. Word articulation scores are
markedly affected by the intensity level of the talker's voice (see

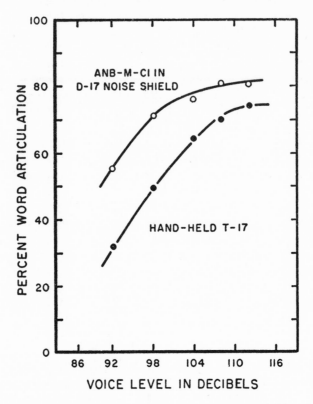

VOICE LEVEL IN DECIBELS

FIG. 49. The relation between word articulation and voice level. The gain was held
constant. Both the talker and listeners were in ambient noise of 120 db.[4] The curves
are for two different microphone installations.

Figure 49). If the purpose of the articulation tests is to compare several
pieces of equipment, it is important that the talker keep his voice level
as nearly constant as possible from test to test. In particular the talker
must not adjust his voice level to compensate for variations in signal
level produced by the equipment itself. Although talkers can learn to
maintain a reasonably steady speaking level through practice, the ex-

perimenter should also provide visual and auditory means of indicating the talker's voice level.

One way of monitoring the talker's voice level is to use the output voltage of the talker's microphone to activate an output meter. This method is satisfactory only if the speech-to-noise ratio is at least 5 db. An additional restriction is that this procedure will not work with many types of carbon microphone. At high voice levels these microphones distort and compress the speech signal. As a result, changes in the output voltage are often not proportional to changes in voice level.

A better monitoring system is one which is independent of the communication system being tested. For example, a magnetic throat microphone with an amplifier and output meter makes a satisfactory visual aid. These microphones provide high speech-to-noise ratios and have an output which is almost exactly proportional to voice level. Additional help to the talker can be provided by introducing into his headset a "side tone" from the throat microphone. This side tone should be adjusted to a high level even though the listeners may be hearing a weak signal. The principal deficiency of this method arises from the difficulty of placing a throat microphone in the same position on the throat from test to test, but variations from this source can be reduced with care.

Allowing talkers to select their speaking levels. For some kinds of tests the experimenter may want to violate the rules given immediately above by allowing the talkers to speak at voice levels which seem appropriate to them. This is usually the case when articulation tests are being made on entire communication systems which already have side tone designed into them. In actual practice the person who uses a communication system tends to adjust the intensity of his voice in accordance with the loudness of the side tone he hears. When the side tone is weak, the talker tends to speak louder; when the side tone is loud, he tends to lower his voice. Such changes in voice level affect not only the level of received speech but also the speech-to-noise ratio provided by the microphone. Thus the efficiency of a communication system depends to a large extent on the voice level selected by the talker, and the voice level selected by the talker depends in turn on the loudness of the side tone he hears. If you want to compare communication systems under realistic conditions, you may want to give the announcers complete freedom to pick the voice levels which seem natural to them with each system they use.

EXPERIMENTAL DESIGN

Articulation test scores must be unbiased if they are to be of any use in comparing different systems. This means that no one system should be tested under more favorable circumstances than those used in testing the other systems. Another way of saying this is that all comparisons of communication devices should be made with conditions as uniform as possible. For example, if the tests involve two microphones,

TABLE 43.　*A sample experimental design for twenty-four articulation tests on three types of microphones. The Arabic numbers show the order in which successive tests are made. The numbers in parentheses identify the word list used in each test*

Day	1				2			
Talker	A	B	C	D	A	B	C	D
Microphone 1	3 (1)	9 (2)	4 (3)	10 (4)	22 (5)	19 (6)	16 (7)	23 (8)
Microphone 2	12 (2)	2 (3)	6 (4)	8 (1)	15 (6)	18 (7)	21 (8)	13 (5)
Microphone 3	5 (3)	7 (4)	11 (1)	1 (2)	17 (7)	20 (8)	24 (5)	14 (6)

the microphones should be tested with the same talkers, same listeners, same amplifiers, same noise, and so on.

In general, the human elements in the test are harder to control than the electronic components. One kind of protection against human bias is the use of a large number of listeners and talkers. Training is another. But even after a crew has undergone considerable training there is still likely to be a slight but rising trend in its scores. In addition, scores may show a slight downward trend during a single day's testing due to fatigue of the listeners or talkers, a change in the characteristics of the noise, or a change in the manner of speaking. Other factors may cause a rise in scores during a single day or from one day to the next.

An experimental design to avoid bias. The best way to guard against bias arising from such unavoidable variability is to design the tests so that the effects of irrelevant factors are distributed randomly over all the tests and equally over all the systems. As a practical matter it is usually convenient to divide the tests into two or more blocks within each of which the testing order is random. Such an experimental design is shown in Table 43. This is for three microphones and for two blocks

TABLE 44. *A testing order for twenty-four articulation tests on three types of microphones. This table is derived from the entries in Table 43*

	Day 1				Day 2		
Test number	Talker	Micro- phone	Word list	Test number	Talker	Micro- phone	Word list
1	D	3	2	13	D	2	5
2	B	2	3	14	D	3	6
3	A	1	1	15	A	2	6
4	C	1	3	16	C	1	7
5	A	3	3	17	A	3	7
6	C	2	4	18	B	2	7
7	B	3	4	19	B	1	6
8	D	2	1	20	B	3	8
9	B	1	2	21	C	2	8
10	D	1	4	22	A	1	5
11	C	3	1	23	D	1	8
12	A	2	2	24	C	3	5

of tests. Each block of trials is run on a separate day and four talkers, A, B, C, and D, participate in the tests.

The test numbers for the first block, 1 to 12, were inserted at random into the columns labelled Day 1; the test numbers for the second block, 13 to 24, were inserted at random in the columns labelled Day 2. This was done by going to a table of random numbers and starting at some haphazard position in the table. To make up Table 43, I started reading down pairs of adjacent columns in the table of random numbers looking for numbers from 01 to 12. The first number I came to was 03. It was entered in the first position under Talker A, Microphone 1. The next number, 09, went under Talker B. The next number was an 04.

This went under Talker C. The next number was another 03, but it was discarded since the 3 had already been used. The next unused number, 10, went under Talker D. And so on.

TABLE 45. *A hypothetical set of articulation test scores obtained on 3 microphones with 4 talkers and 6 listeners. The testing order is shown in Table 44*

		Day 1				Day 2			
	Talkers →	A	B	C	D	A	B	C	D
Listeners									
Microphone 1	E	80	92	86	82	98	86	44	76
	F	82	60	68	70	86	88	58	74
	G	48	68	62	64	86	82	46	70
	H	80	86	62	60	84	70	72	88
	I	74	86	80	76	78	76	74	68
	J	54	66	70	82	58	86	74	56
Microphone 2	E	86	82	74	48	88	76	62	48
	F	66	78	44	60	74	86	74	52
	G	36	54	42	38	74	50	72	60
	H	82	56	66	44	80	64	44	50
	I	46	80	38	74	52	68	52	60
	J	46	62	40	40	40	52	70	70
Microphone 3	E	42	50	74	46	72	60	42	42
	F	70	68	40	34	50	68	56	42
	G	68	48	48	64	48	46	56	44
	H	72	54	62	72	46	66	38	60
	I	74	80	64	72	52	72	36	42
	J	44	54	50	46	74	54	40	40

Notice that if the experimenter wanted to use a third block of trials he could add it to the table and insert randomly into that block the numbers from 25 to 36.

Assignment of the word lists. Table 43 shows that eight word lists have been used in this experimental plan but that they have not been assigned at random. The word lists have been systematically varied so that each microphone is tested once and only once with each word list. In addition each talker uses each word list only once.

A testing order. From the experimental plan in Table 43 the experimenter can now arrange a testing order for the 24 tests. This is shown in Table 44. The first line of Table 44 means that the first test of the series is made with Talker D using Microphone 3 and word list 2. The second test is made with Talker B, using Microphone 2 and word list 3. And so on.

TABLE 46. *The number of articulation scores (N) which can be used to evaluate each variable, and the mean articulation score (M) for each of the principal variables in Table 45*

Variable	N	M
Microphone 1	48	73.2
Microphone 2	48	60.4
Microphone 3	48	55.0
Talker A	36	66.4
Talker B	36	68.7
Talker C	36	57.8
Talker D	36	58.7
Day 1	72	62.7
Day 2	72	63.1
Listener E	24	68.2
Listener F	24	64.5
Listener G	24	57.2
Listener H	24	64.9
Listener I	24	65.6
Listener J	24	57.0

Sample results. Not discussed above is the fact that 6 listeners are used throughout the entire series of tests. These are the individuals, E, F, G, H, I, and J, in Table 45. Table 45 shows a hypothetical set of articulation test scores obtained from this experiment. For example, the six listeners got articulation scores of 46, 34, 64, 72, 72, and 46, respectively, in the first test of the series—that made with Talker D using Microphone 3.

By combining the scores in Table 45 in different ways, we can come up with several interesting comparisons. Adding the 48 scores for

TABLE 47.   *Analysis of variance of the articulation test scores in Table 45*

Source of variance	Sum of squares	Degrees of freedom	Estimate of variance	F-ratio
Between microphones (M)	8,402.06	2	4,201.03	$4,201.03/145.03 = 28.97^*$
Between days (D)	4.69	1	4.69	$4.69/153.78 = 0.03$
Between talkers (T)	3,231.42	3	1,077.14	$1,077.14/153.78 = 7.00^{**}$
Between listeners (L)	2,599.14	5	519.83	$519.83/153.78 = 3.38^{***}$
Interactions: M × D	882.72	2	441.36	$441.36/145.03 = 3.04$
M × T	304.16	6	50.69	$50.69/145.03 = 0.35$
M × L	1,364.94	10	136.49	$136.49/145.03 = 0.94$
D × T	364.09	3	121.36	$121.36/153.78 = 0.79$
D × L	1,391.81	5	278.36	$278.36/153.78 = 1.81$
T × L	2,706.08	15	180.41	$180.41/153.78 = 1.17$
M × D × T	1,096.8c	6	182.80	$182.80/145.03 = 1.26$
M × D × L	1,172.28	10	117.23	$117.23/145.03 = 0.81$
M × T × L	4,268.84	30	142.29	$142.29/145.03 = 0.98$
D × T × L	2,306.74	15	153.78	
M × D × T × L	4,350.84	30	145.03	
Total	34,446.64	143		

* Significant at the 0.001 level, that is, $P < 0.001$.
** Significant at the 0.01 level, that is, $0.001 < P < 0.01$.
*** Significant at the 0.05 level, that is, $0.01 < P < 0.05$.

Microphone 1 and dividing by 48 gives an average articulation score for this microphone of 73.2 (see Table 46). A similar computation for Microphone 2 gives an articulation score of 60.4. For Microphone 3 the average score is only 55.0.

Combining the scores in a different way shows that there were some differences among the talkers. Average articulation scores were highest when Talker B did the talking, and lowest when Talker C took over.

The average scores for Day 2 were somewhat higher than for Day 1. Finally, the listeners also differed in their average scores. Listener E got the highest scores; J the lowest.

The analysis of variance of the data. To evaluate the significance of the average differences found above, the data in Table 45 should be subjected to an analysis of variance—the results of which appear in Table 47. This analysis shows that the differences between microphones are highly significant. The differences between talkers are also highly significant, and the differences among listeners are significant but not highly so. The difference between days is not significant, which tells us that the small rise between the two days is small enough to be due to chance. Since none of the interactions is significant, we can feel confident that the differences between microphones are consistent for different talkers, for different listeners, and are repeatable from day to day.

Although this analysis is a bit more complicated than anything you have had before, it follows exactly the same rules as were developed in Chapters 4 and 5. As an exercise start with the data in Table 45 and see if you can come out with the results in Table 47.

Bibliography

1. Adams, J. K.: *Basic Statistical Concepts.* New York: McGraw-Hill, 1955.
2. American Standards Association (70 East Forty-fifth Street, New York 17, N.Y.): American recommended practice for compiling industrial accident causes. American Standard Z16.2–1941, UDC 313.1. August 1, 1941.
3. Andrews, T. G. (ed.): *Methods of Psychology.* New York: Wiley, 1948.
4. Anonymous: Articulation testing methods II. Office of Scientific Research and Development, National Defense Research Committee, Division 17, Section 17.3, Contract OEMsr–658 (Psycho-Acoustic Laboratory, Harvard University, Cambridge, Mass.), OSRD Report No. 3802. November 1, 1944.
5. Anonymous: Motion and time analysis of A/A coaching and CIC layout: CL–89–USS Miami. Systems Research Laboratory (Cruft Building, Harvard University, Cambridge 38, Mass., Service Control No. NS–343, Contract OEMsr–658), Research Report No. 3. October 30, 1945.
6. Anonymous: Motion and time analysis of combat information centers: Tucson, Nashville, Louisville, Boston. Systems Research Laboratory (Cruft Building, Harvard University, Cambridge 38, Mass., Service Control No. NS–343, Contract OEMsr–658), Research Report No. 4. November 1, 1945.
7. Anonymous: A study of plotting techniques. Systems Research Laboratory (Cruft Building, Harvard University, Cambridge 38, Mass., Service Control No. NS–343, Contract OEMsr–658), Research Report No. 11. December 24, 1945.
8. Ashby, R. M., Josephson, V., and Sydoriak, S.: Signal threshold studies. Office of Naval Research, Naval Research Laboratory, Washington, D.C., Report No. R–3007. December 1, 1946.
9. Barnes, R. M.: *Motion and Time Study.* (3rd ed.) New York: Wiley, 1949.
10. Bartlett, N. R., and Sweet, A. L.: Visibility on cathode-ray tube screens: signals on a P–7 screen exposed for different intervals. *Journal of the Optical Society of America,* **39** (June, 1949), 470–473.
11. Beranek, L. L.: *Acoustic Measurements.* New York: Wiley, 1949.
12. Bessey, E. G., and Machen, G. S.: An operational test of laboratory determined optima of screen brightness and ambient illumination for radar reporting rooms. *Journal of Applied Psychology,* **41** (February, 1957), 51–52.

13. Blackwell, H. R.: Psychophysical thresholds: experimental studies of methods of measurement. Engineering Research Institute, University of Michigan (Ann Arbor, Mich.), Engineering Research Bulletin No. 36. January, 1953.
14. Bridgman, P. W.: The prospect for intelligence. *The Yale Review*, **34** (March, 1945), 444–461.
15. Bridgman, P. W.: *The Logic of Modern Physics*. New York: Macmillan, 1948.
16. Brower, D.: The role of incentive in psychological research. *Journal of General Psychology*, **39** (July, 1948), 145–147.
17. Brown, C. W., and Ghiselli, E. E.: *Scientific Method in Psychology*. New York: McGraw-Hill, 1955.
18. Channell, R. C.: An analysis of pilots' performances in multi-engine aircraft (R5D). Division of Bio-Mechanics, The Psychological Corporation, New York, U. S. Navy Special Devices Center, Contract N6ori–151, Task Order No. 1, Project No. 20–0–1. April 15, 1947.
19. Chapanis, A.: Evaluation of the "Contrast Discrimination Test" as an anoxia demonstration device. Army Air Forces Air Technical Service Command, Aero Medical Laboratory (Wright Field, Dayton, Ohio), Memorandum Report TSEAA–695–65. March 1, 1946.
20. Chapanis, A.: The dark adaptation of the color anomalous. *The American Journal of Physiology*, **146** (August, 1946), 689–701.
21. Chapanis, A.: Speed of reading target information from a direct-reading, counter-type indicator versus conventional radar bearing-and-range dials. Psychological Laboratory, The Johns Hopkins University (Baltimore, Md.), Report No. 166–I–3. November 1, 1946.
22. Chapanis, A.: The relative efficiency of a bearing counter and bearing dial for use with PPI presentations. Psychological Laboratory, The Johns Hopkins University (Baltimore, Md.), Report No. 166–I–26, August 1, 1947.
23. Chapanis, A.: Some aspects of operator performance on the VJ remote radar indicator. Psychological Laboratory, The Johns Hopkins University (Baltimore, Md.). Office of Naval Research, Special Devices Center, Port Washington, L.I., N.Y., Technical Report SDC 166–I–91. June 20, 1949.
24. Chapanis, A.: How we see: a summary of basic principles. Chap. 1, pp. 3–60 in Panel on Psychology and Physiology, Committee on Undersea Warfare: *A Survey Report on Human Factors in Undersea Warfare*. Washington, D.C.: National Research Council, 1949.
25. Chapanis, A.: Theory and methods for analyzing errors in man-machine systems. *Annals of the New York Academy of Sciences*, **51** (January, 1951), 1179–1203.
26. Chapanis, A.: Random-number guessing behavior. *American Psychologist*, **8** (August, 1953), 332.
27. Chapanis, A.: *The Design and Conduct of Human Engineering Studies*. San Diego, Calif.: San Diego State College Foundation, 1956.
28. Chapanis, A.: L'adaptation de la machine à l'homme: l'étude des relations homme-machine. *Revue de Psychologie Appliquée*, **6** (October, 1956), 213–234.
29. Chapanis, A., Garner, W. R., and Morgan, C. T.: *Applied Experimental Psychology: Human Factors in Engineering Design*. New York: Wiley, 1949.
30. Chapanis, A., Garner, W. R., Morgan, C. T., and Sanford, F. H.: Lectures on men and machines: an introduction to human engineering. Department of

Psychology, The Johns Hopkins University (Baltimore, Md.), Report No. 166–I–19. 1947. 246 pages.

31. Christensen, J. M.: The sampling method of activity analysis and its application to the problem of aircraft crew requirements. Pp. 37–54 in Morrison, D. (ed.): *Scientific Methods for Use in the Investigation of Flight Crew Requirements.* Flight Safety Foundation (Woods Hole, Mass.), Project RP–1–F. November, 1948.

32. Christensen, J. M.: Arctic aerial navigation: a method for the analysis of complex activities and its application to the job of the arctic aerial navigator. *Mechanical Engineering,* **71** (January, 1949), 11–16 and 22.

33. Christensen, J. M.: A sampling technique for use in activity analysis. *Personnel Psychology,* **3** (Autumn, 1950), 361–368.

34. Churchill, A. V.: Comparison of two visual display presentations. *Journal of Applied Psychology,* **40** (April, 1956), 135.

35. Churchman, C. W.: *Theory of Experimental Inference.* New York: Macmillan, 1948.

36. Cochran, W. G., and Cox, Gertrude M.: *Experimental Designs.* (2nd ed.) New York: Wiley, 1957.

37. Cohen, M. R., and Nagel, E.: *An Introduction to Logic and Scientific Method.* New York: Harcourt, Brace, 1934.

38. Craik, K. J. W., and Vernon, M. D.: The nature of dark adaptation. *British Journal of Psychology,* **32** (July, 1941), 62–81.

39. David, E. E., Jr.: Naturalness and distortion in speech-processing devices. *Journal of the Acoustical Society of America,* **28** (July, 1956), 586–589.

40. Deming, W. E.: *Some Theory of Sampling.* New York: Wiley, 1950.

41. Dixon, W. J., and Massey, F. J.: *Introduction to Statistical Analysis.* New York: McGraw-Hill, 1951.

42. Doughty, J. M.: The effect of psychophysical method and context on pitch and loudness functions. *Journal of Experimental Psychology,* **39** (October, 1949), 729–745.

43. Dunlap, J. W.: Men and machines. *Journal of Applied Psychology,* **31** (December, 1947), 565–579.

44. Edwards, A. L.: *Experimental Design in Psychological Research.* New York: Rinehart, 1950.

45. Edwards, A. L.: *Statistical Methods for the Behavioral Sciences.* New York: Rinehart, 1954.

46. Egan, J. P., Carterette, E. C., and Thwing, E. J.: Some factors affecting multichannel listening. *Journal of the Acoustical Society of America,* **26** (September, 1954), 774–782.

47. Faber, S.: Ground-simulator study of the effects of stick force and displacement on tracking performance. Langley Aeronautical Laboratory (Langley Field, Va.) National Advisory Committee for Aeronautics, Technical Note 3428. April, 1955.

48. Fisher, R. A., and Yates, F.: *Statistical Tables for Biological, Agricultural and Medical Research.* London: Oliver and Boyd, 1949.

49. Fitts, P. M. (ed.): *Human Engineering for an Effective Air-Navigation and Traffic-Control System.* Washington, D.C.: National Research Council, 1951.

50. Fitts, P. M., and Jones, R. E.: Analysis of factors contributing to 460 "pilot-

error" experiences in operating aircraft controls. Army Air Forces Air Materiel Command, Engineering Division, Aero Medical Laboratory (Wright-Patterson Air Force Base, Ohio), Report No. TSEAA–694–12. July 1, 1947.

51. Fitts, P. M., and Jones, R. E.: Psychological aspects of instrument display. I. Analysis of 270 "pilot-error" experiences in reading and interpreting aircraft instruments. U. S. Air Forces Air Materiel Command, Engineering Division, Aero Medical Laboratory (Wright-Patterson Air Force Base, Ohio), Report No. TSEAA–694–12A. October 1, 1947.

52. Fitts, P. M., Jones, R. E., and Milton, J. L.: Eye movements of aircraft pilots during instrument-landing approaches. *Aeronautical Engineering Review,* 9 (February, 1950), 24–29.

53. Flanagan, J. C.: The critical incident technique. *Psychological Bulletin,* 51 (July, 1954), 327–358.

54. Flexman, R. E., Matheny, W. G., and Brown, E. L.: Evaluation of the school Link and special methods of instruction in a ten-hour private pilot flight-training program. University of Illinois, The Institute of Aviation (Urbana, Ill.), Bulletin Number 80 (Aeronautics Bulletin Number 8), Vol. 47, July, 1950.

55. Fraser, D. C.: The relation of an environmental variable to performance in a prolonged visual task. *Quarterly Journal of Experimental Psychology,* 5 (February, 1953), 31–32.

56. Fryer, H. C.: *Elements of Statistics.* New York: Wiley, 1954.

57. Gardner, J. F., and Lacey, R. J.: An experimental comparison of five different attitude indicators. Aero Medical Laboratory, Wright Air Development Center (Wright-Patterson Air Force Base, Ohio), WADC Technical Report 54–32. May, 1954.

58. Gebhard, J. W.: Some experiments with the VF aided tracking equipment. Psychological Laboratory, The Johns Hopkins University (Baltimore, Md.), Report No. 166–I–53. September 15, 1948.

59. Geldard, F. A.: *The Human Senses.* New York: Wiley, 1953.

60. Gibbs, C. B., and Brown, I. D.: Increased production from the information incentive in a repetitive task. Medical Research Council, Applied Psychology Research Unit (15 Chaucer Road, Cambridge, England), Report No. APU 230. March, 1955.

61. Grether, W. F.: Instrument reading. I. The design of long-scale indicators for speed and accuracy of quantitative readings. *Journal of Applied Psychology,* 33 (August, 1949), 363–372.

62. Guilford, J. P.: *Psychometric Methods.* (2nd ed.) New York: McGraw-Hill, 1954.

63. Heinrich, H. W.: *Industrial Accident Prevention.* New York: McGraw-Hill, 1950.

64. Heiland, R. E., and Richardson, W. J.: *Work Sampling.* New York: McGraw-Hill, 1957.

65. Hoffman, C. E. (Chairman): *Organization and Administration of the Military Research and Development Programs: Twenty-fourth Intermediate Report of the Committee on Government Operations.* (Union Calendar No. 895. House Report No. 2618). Washington, D.C.: United States Government Printing Office, 1954.

66. Jenkins, W. L., and Connor, Minna B.: Some design factors in making settings on a linear scale. *Journal of Applied Psychology,* 33 (August, 1949), 395–409.

67. Kappauf, W. E.: Design of instrument dials for maximum legibility: I. Development of methodology and some preliminary results. U. S. Air Forces Air Materiel Command, Aero Medical Laboratory, Engineering Division (Wright-Patterson Air Force Base, Ohio), Memorandum Report No. TSEAA–694–1L. October 20, 1947.

68. Kempthorne, O.: *The Design and Analysis of Experiments*. New York: Wiley, 1952.

69. Lasagna, L., and Von Felsinger, J. M.: The volunteer subject in research. *Science*, **120** (September 3, 1954), 359–361.

70. Lauer, A. R., and McMonagle, J. C.: Do road signs affect accidents? *Traffic Quarterly*, **9** (July, 1955), 322–329.

71. LeShan, L. L., and Brame, J. B.: A note on techniques in the investigation of accident prone behavior. *Journal of Applied Psychology*, **37** (April, 1953), 79–81.

72. Leyzorek, M.: Mounting angle of a VJ remote radar indicator and its effect on operator performance. Psychological Laboratory, The Johns Hopkins University (Baltimore, Md.), Report No. 166–I–41, February 10, 1948.

73. Lindquist, E. F.: *Design and Analysis of Experiments in Psychology and Education*. New York: Houghton Mifflin, 1953.

74. Loucks, R. B.: Evaluation of aircraft attitude indicators on the basis of Link Instrument Ground Trainer performance. 27th AAF Base Unit, AAF School of Aviation Medicine (Randolph Field, Tex.), Project No. 341, Report No. 1. June 22, 1945.

75. Loucks, R. B.: An experimental evaluation of the interpretability of various types of aircraft attitude indicators. Pp. 111–135 in Fitts, P.M. (ed.): *Psychological Research on Equipment Design*. Washington, D.C.: Government Printing Office, 1947.

76. Lutz, Mary C., and Chapanis, A.: Expected locations of digits and letters on ten-button keysets. *Journal of Applied Psychology*, **39** (October, 1955), 314–317.

77. McCollom, I. N., and Chapanis, A.: *A Human Engineering Bibliography*. San Diego, Calif.: San Diego State College Foundation, 1956.

78. McCormick, E. J.: *Human Engineering*. New York: McGraw-Hill, 1957.

79. McFarland, R. A., Moore, R. C., and Warren, A. B.: *Human Variables in Motor Vehicle Accidents*. Harvard School of Public Health (One Shattuck Street), Boston, Mass. 1955.

80. McFarland, R. A., and Moseley, A. L.: *Human Factors in Highway Transport Safety*. Harvard School of Public Health (695 Huntington Avenue), Boston, Mass. 1954.

81. McNemar, Q.: *Psychological Statistics*. (2nd ed.) New York: Wiley, 1955.

82. Melville, G. W.: The engineer and the problem of aerial navigation. *The North American Review*, **173** (December, 1901), 820–831.

83. Mowrer, O. H., Rayman, N. N., and Bliss, E. L.: Preparatory set (expectancy)— an experimental demonstration of its "central" locus. *Journal of Experimental Psychology*, **26** (April, 1940), 357–372.

84. Mundel, M. E.: Motion study techniques which could be brought to bear on desirable size of aircraft crews. Pp. 55–68 in Morrison, D. (ed.): *Scientific Methods for Use in the Investigation of Flight Crew Requirements*. Flight

Safety Foundation (Woods Hole, Mass.), Project RP–1–F. November,
1948.

85. Mundel, M. E.: *Motion and Time Study: Principles and Practice.* (2nd ed.)
New York: Prentice-Hall, 1955.

86. Office of Scientific Research and Development: *Transmission and Reception of
Sounds under Combat Conditions.* Summary Technical Report of Division 17,
NDRC. Vol. 3. Washington, D.C., 1946.

87. O'Neil, W. M.: *An Introduction to Method in Psychology.* Melbourne, Australia:
Melbourne University Press, 1957.

88. Payne, S. L.: *The Art of Asking Questions.* Princeton, N.J.: Princeton University
Press, 1951.

89. Peters, C. C., and Van Voorhis, W. R.: *Statistical Procedures and Their Mathematical Bases.* New York: McGraw-Hill, 1940.

90. Pincus, G., and Hoagland, H.: Effects on industrial production of the administration of Δ5 pregnenolone to factory workers, I. *Psychosomatic Medicine,*
7 (November, 1945), 342–346.

91. Pollock, W. T., and Chapanis, A.: The apparent length of a line as a function
of its inclination. *Quarterly Journal of Experimental Psychology,* **4** (November, 1952), 170–178.

92. Randall, F. E., Damon, A., Benton, R. S., and Patt, D. I.: Human body size in
military aircraft and personal equipment. Army Air Forces Air Materiel
Command (Wright Field, Dayton, Ohio), Technical Report No. 5501.
June 10, 1946.

93. Rand Corporation: *A Million Random Digits with 100,000 Normal Deviates.*
Glencoe, Illinois: The Free Press, 1955.

94. Riggs, Margaret M., and Kaess, W.: Personality differences between volunteers
and non-volunteers. *Journal of Psychology,* **40** (October, 1955), 229–245.

95. Roethlisberger, F. J., and Dickson, W. J. *Management and the Worker.* Cambridge, Mass.: Harvard University Press, 1939.

96. Roy, R. H.: Do wage incentives reduce costs? *Industrial and Labor Relations
Review,* **5** (January, 1952), 195–208.

97. Scales, Edythe M., and Chapanis, A.: The effect on performance of tilting the
toll-operator's keyset. *Journal of Applied Psychology,* **38** (December, 1954),
452–456.

98. Senders, Virginia L.: Further analysis of response sequences in the setting of a
psychophysical experiment. *American Journal of Psychology,* **66** (April,
1953), 215–228.

99. Senders, Virginia L., and Cohen, J.: The influence of methodology on research
on instrument displays. Wright Air Development Center, Air Research and
Development Command (Wright-Patterson Air Force Base, Ohio), WADC
Technical Report 53–93. April, 1953.

100. Senders, Virginia L., and Sowards, Ann: Analysis of response sequences in the
setting of a psychophysical experiment. *American Journal of Psychology,* **65**
(July, 1952), 358–374.

101. Siegel, S.: *Nonparametric Statistics for the Behavioral Sciences.* New York:
McGraw-Hill, 1956.

102. Sinaiko, H. W., and Buckley, E. P.: Human factors in the design of systems.
Naval Research Laboratory (Washington, D.C.), Report No. 4996. August
29, 1957.

103. Smith, A. A., and Boyes, G. E.: Visibility on radar screens: the effect of CRT bias and ambient illumination. *Journal of Applied Psychology,* **41** (February, 1957), 15–18.

104. Sullivan, M.: *Our Times: 1900–1925.* Vol. II. New York: Charles Scribner's Sons, 1936.

105. Tinker, M. A.: Effect of visual adaptation upon intensity of light preferred for reading. *American Journal of Psychology,* **54** (October, 1941), 559–563.

106. Townsend, J. C.: *Introduction to Experimental Method for Psychology and the Social Sciences.* New York: McGraw-Hill, 1953.

107. Tufts College, Institute of Applied Experimental Psychology: *Handbook of Human Engineering Data.* (2nd ed.) Office of Naval Research, Special Devices Center, NavExos P–643, Technical Report No. SDC 199–I–2. 1952.

108. Vasilas, J. N., Fitzpatrick, R., Dubois, P. H., and Youtz, R. P.: Human factors in near accidents. Air University, USAF School of Aviation Medicine (Randolph Field, Tex.), Report No. 1, Project No. 21–1207–0001. June, 1953.

109. Verplanck, W. S., Collier, G. H., and Cotton, J. W.: Nonindependence of successive responses in measurements of the visual threshold. *Journal of Experimental Psychology,* **44** (October, 1952), 273–282.

110. Verplanck, W. S., Cotton, J. W., and Collier, G. H.: Previous training as a determinant of response dependency at the threshold. *Journal of Experimental Psychology,* **46** (July, 1953), 10–14.

111. Von Neumann, J.: The general and logical theory of automata. Pp. 1–31 in Jeffress, L. A. (ed.): *Cerebral Mechanisms in Behavior.* New York: Wiley, 1951.

112. Weldon, R. J., and Peterson, G. M.: Factors influencing dial operation: three-digit multiple-turn dials. Sandia Corporation (Albuquerque, N.M.), Engineering Research Report SC–3659(TR). February 28, 1955.

113. Williams, S. B.: Visibility on radar scopes. Chap. 4, Pp. 101–130 in Panel on Psychology and Physiology, Committee on Undersea Warfare: *A Survey Report on Human Factors in Undersea Warfare.* Washington, D.C.: National Research Council, 1949.

114. Williams, S. B., Bartlett, N. R., and King, E.: Visibility on cathode-ray tube screens: screen brightness. *Journal of Psychology,* **25** (April, 1948), 455–466.

115. Williams, S. B., and Hanes, R. M.: Visibility on cathode-ray tube screens: intensity and color of ambient illumination. *Journal of Psychology,* **27** (January, 1949), 231–244.

116. Wilson, E. B.: *An Introduction to Scientific Research.* New York: McGraw-Hill, 1952.

117. Woodson, W. E.: *Human Engineering Guide for Equipment Designers.* Berkeley and Los Angeles: University of California Press, 1954.

118. Yates, F.: *Sampling Methods for Censuses and Surveys.* New York: Hafner Publishing Company, 1949.

Index

Carterette, E. C., 216, 274, 305
Causes, study of, in human behavior, 15, 25, 71–72, 94–95, 149–150
Central tendency, measures of, 104–107
Chance effects, in evaluating experimental outcomes, 121, 126, 142–144
Changes during an experiment, controlling for, 153–156, 159, 183–184, 192-197, 230–234, 296–298
Channell, R. C., 44–46, 50, 52–53, 304
Chapanis, A., 2, 7–8, 113–114, 130, 132, 136–137, 146, 158, 178, 194, 204, 217, 219, 223, 227–228, 235, 237, 240–241, 248–249, 275, 304, 307, 308
Check lists, 17
Christensen, J. M., 27–29, 33–34, 36, 305
Churchill, A. V., 205, 305
Churchman, C. W., 3, 305
CIC (combat information center), studies of, 17, 53–60, 219
Classes and class limits in frequency distributions, 97–99
Cochran, W. G., 198, 305
Cockpit, aircraft, study of, 44–46, 50–52
instruments, study of, 61–62
Cohen, J., 217, 308
Cohen, M. R., 3, 305
Collier, G. H., 309
Colorblindness, operational definition, 20
Common sense vs. research, 5–10
Communication in noise, 7, 216–217, 279–280
Communication systems, tests of, see Articulation tests
Comparisons of groups, experimental designs for, 157–165
Complexity of man, 11–12, 15–16
Concepts, operational definition of, 18–21
Confidence limits, 145–146
Confounding variables, 71–75, 156–157
Connor, M. B., 212, 306
Consciousness, as object of study, 15
Constants, method of, 263–270
Context effects in psychophysics, 267–268
Control observations, methods of collecting, 233–234
need for, 230–233
Controls in human studies, 199–203
designing, 148–149
illustration of, 177–179
need for, 73–75, 230–233
relation to sensitivity of experiments, 163–165, 201–203
Controls, machine, studies of, 211–213

Correlation, 113–120
Pearson product-moment coefficient of, 115–118
partial, 118–120
Cotton, J. W., 309
Counterbalancing trials to avoid bias, 153–155, 160
illustration of, 182–184, 296–298
by means of Latin-square, 197–198
Cox, G. M., 198, 305
Craik, K. J. W., 259–260, 305
Criteria for evaluating layouts, 59–60
Criteria for matching subjects, 159, 162
Criterion measures in human engineering experiments, 215–220
Critical-incident technique, 88–92
Crowding, index of, 60
Cumulative frequency distribution, 101–104, 106–107

Damon, A., 100, 308
Dark adaptation, experiment, 177–179
variability in measurements, 258–259
Data sheets, for activity analyses, 27–33
general, 17
see also Accident report forms
David, E. E., Jr., 216, 305
Decrement during an experiment, controlling for, 153–156, 159, 183–184, 192–197, 230–234, 296–298
Degrees of freedom, definition, 124–125
use of, 128–129, 131–132, 134, 136, 138, 161–162, 167, 169, 173–175, 300–301
Δ, 256
Deming, W. E., 305
Dials, study of, 203–205, 217–18, 225–28
Dickson, W. J., 25, 74, 308
Difficulties in studying people, 3–4, 11–15, 73–75, 96–97
Distribution, frequency see Frequency distribution
Dixon, W. J., 305
DL, see Threshold, differential
Doughty, J. M., 267–268, 305
Driving, bus, activity analysis of, 29–30
truck, near-accident study of, 88
Drugs, experiment on, 230–233
Dubois, P. H., 86–88, 309
Dummies, use of in human engineering studies, 18
Dunlap, J. W., 206, 245, 305

Earplugs and communication in noise, 7
Edwards, A. L., 305
Egan, J. P., 216, 274, 305

313

Lacey, R. J., 244–245, 306
Langley, S. P., 6
Lasagna, L., 307
Latin-square experimental design, 192–198
Lauer, A. R., 120, 307
Layout of work and equipment, study of, 39–62
Learning during an experiment, controlling for, 153–156, 159, 183–184, 192–197, 230–234, 251–252, 296–298
LeShan, L. L., 79–80, 94, 307
Level of significance, 125–127
Leyzorek, M., 197, 219, 307
Librarian's job, activity sampling of, 26–36
Lighting, studies of, 219–220, 232–233
Limen, see Threshold
Limits, method of, 261–263
Lindenbaum, L. E., 9–10
Lindquist, E. F., 198, 307
Link analysis, 51–62
Link trainer, study of, 123–124, 158
Listeners for articulation tests, 281–283
Loucks, R. B., 8, 91, 219, 307
Lutz, M. C., 240–241, 307

Machen, G. S., 205, 303
Man, complexity of vs. machine, 11–12, 15–16
Masking effect of noise, 261–262
Massey, F. J., 305
Matched-groups design, 159–160
Matched subjects design, 160
Matheny, W. G., 123, 306
McCollom, I. N., 235, 307
McCormick, E. J., 2, 307
McCulloch, W., 12
McFarland, R. A., 24, 29–30, 88, 94–95, 307
McMonagle, J. C., 120, 307
McNemar, Q., 307
Mean, 104–107
Mean difference, significance of, 122–124, 127–131, 160–162
Median, 106–107
Melville, G. W., 6, 307
Memomotion study, 36–37
Mental images, as objects of study, 15, 19–21
Method of adjustment, 260–261
Method of constants, 263–270
Method of limits, 261–263
Micromotion study, 62–71

Microphones, experimental design for testing, 296–301
tests of, 279, 283–284
Milton, J. L., 61, 306
Mind, as object of study, 15, 19–21
Models for evaluating layouts, 59
Monosyllabic words for articulation tests, 286–287
Moore, R. C., 307
Morgan, C. T., 2, 7–8, 217, 227–228, 304
Moseley, A. L., 24, 29–30, 88, 94–95, 307
Motion picture camera, in micromotion study, 62
in observation, 36–37, 48, 61
Motion study, 23–24, 62–71
Motives, study of, 15
Motivation, controlling, 221–224, 282
difficulty of, 205–207
Movements, hand, classification of, 62–71
in driving, 29–30
Mowrer, O. H., 225–226, 307
Multiple-process charts, see Process charts
Multi-variable experiments, 165–176, 179–198, 296–301
Mundel, M. E., 47–49, 64–69, 307, 308

Nagel, E., 3, 305
Naive subjects, use of, 244–246, 281
Navigation, aerial, see Aerial navigation
Near-accidents, Air Force study of, 86–88
general, 85–88, 92–95
in long-haul truck operations, 88
limitations of data, 92–95
methods of collecting data on, 86–88
Newcomb, S., 6
Newspaper advertisements, study of, 37–38
Noise, communication in, 7, 216–217, 279–280
Nonsense syllables, for articulation tests, 286
Normal distribution, 100–101, 109
Number of subjects, 236–238
as a factor in increasing sensitivity of tests, 163
in articulation tests, 283

Objectivity in studying man, 13–14, 17–18
Observation, categories of, for activity analyses, 30–31
difficulties of, 12–14, 71–75, 79–80
general rules for making, 16–18, 251

human vs. motion picture camera, 36–37
role of in experimentation, 150
and sensory defects, 13
Observational methods, critique of, 71–75
O'Neil, W. M., 3, 308
Operationism and the study of behavior, 18–21, 214–215
applied to activity analyses, 31
Opinion data, 24–26
Oxygen lack at high altitudes, *see* Anoxia

Parachutes, study of, 157
Partial correlation, 118–119
Patt, D. I., 100, 308
Pay as motivator, 221–222
Payne, S. L., 26, 308
PB (Phonetically-balanced) words, for articulation tests, 286–287
Pearson product-moment coefficient of correlation, 115–116
Percentiles, 112–113
Personnel, *see* Subjects
Peters, C. C., 308
Peterson, G. M., 162, 205, 309
Pilot error, study of, 89–92
Pilot training, study of, 123–124
Pincus, G., 231–233, 308
Pollock, W. T., 275, 308
Practical vs. statistical significance, 144–146
Precision, of apparatus, 163–165, 247–249
of measurements and sensitivity of experiments, 163–165
Preferences as criteria, 218–220
Probability, as level of significance, 125–127
Process analysis, 39–62
Process charts, basic symbols for, 40–43
illustration of, 42
multiple-process charts, 45–52
Product-moment coefficient of correlation, 115–116
Progressive changes during an experiment, controlling for, 153–156, 159, 183–184, 192–197, 230–234, 296–298
Punishment as motivator, 224

Questions, art of asking, 25–26

Radar, instruments, studies of, 204–205
observing, studies of, 27–37
plotting, 63–70
rate-aiding equipment for, 179–192

studies of, 17, 113–115, 197, 205, 219, 247–249
Randall, F. E., 100, 308
Randomizing, assignment of subjects, 158–159
trials to avoid bias, 155, 158–159
illustration of, 183–184, 296–298
word lists in articulation tests, 291
Random sampling, 239–240
Random variability, in evaluating experimental outcomes, 121, 126, 201–203
in experimental design, 146, 151–153, 236–238
Range, 108
Rate-aiding equipment for radar, study of, 179–192
Rayman, N. N., 225–226, 307
Reaction time data, 97–100
Reaction time, study of, 225–226
Record sheets, 17
Regression line, 118
Relationship, measures of, 113–120
Relay assembly study, 73–75
Replication, 136–138, 152–153
Research, common sense versus, 5–10
difficulties in doing human, 11–15
difficulties in teaching, 3–5
general strategy for doing, 15–21
Rest-pauses, study of, 74–75
Richardson, W. J., 34, 306
Riggs, M. M., 308
Roethlisberger, F. J., 25, 74, 308
Roy, R. H., 206, 245, 308

Sampling, accident, 78–79
equipment, 248–250
job, 48–52
in activity analyses, 35–36
situational, 251–252
subject, 238–246
in activity analyses, 35
Sampling duration, for activity analyses, 32–34
Sampling interval, for activity analyses, 34–35
Sanford, F. H., 304
Sawing operation, study of, 47–49
Scales, E. M., 194, 219, 237, 308
Scientific method, difficulties in teaching, 3–5
Senders, V. L., 217, 277, 308
Sensation, as object of study, 15, 19–21
definition of, 254
Sensitivity of experiments, methods of increasing, 163–165, 173–175, 201–203

Sentences, for articulation tests, 288–290
Siegel, S., 308
Significance, level of, 125–127
 practical versus statistical, 144–146
Simo-chart, 63, 70
Sinaiko, H. W., 219, 308
Situational sampling, 251–252
Skewness (*also* Skewed distributions),
 100–101, 107
Sleeping bags, study of, 156–157
Smith, A. A., 309
Sowards, A., 277, 308
Speech intelligibility, tests of, *see* Articu-
 lation tests
Spondaic words, for articulation tests,
 287–288
Standard deviation, 108–112
Standard error, of difference between
 means, 122
 of estimate, 118
Stimulus, 254
Stoves, experiment on the design of,
 9–10, 97–99, 130–131, 136–141
Stratified sampling, 240–242
Subjects, sampling, 35, 238–243
 selection of, 235–246
 for articulation tests, 281–283
Subjects-by-experimental conditions ex-
 perimental design, 131–141, 179–
 192
Sullivan, M., 309
Sum-of-squares, definition of, 112
 use of, 124, 127–135, 138–141, 168
Sweet, A. L., 263, 303
Sydoriak, S., 303
Systematic bias in experiments, 153–155
 counterbalancing to avoid, 160, 182–
 184, 296–298

Talkers for articulation tests, 281–283
Telephone operator's job, activity sampl-
 ing of, 37–39
Telephone operator's keyset, study of,
 192–197
Television pictures, experiment on, 264–
 267
Therbligs, definition of, 62–63
 table of, 64–69
Threshold, absolute, 255
 differential, 255–256
Thwing, E. J., 216, 274, 305
Time and motion study, 23–24, 62–71
Tinker, M. A., 219, 309
Townsend, J. C., 3, 309
Training of subjects, for articulation
 tests, 282–286

for experiments, 181, 244–245
Treatments-by-subjects experimental de-
 sign, *see* Subjects-by-experimental
 conditions experimental design
Truck driving, near-accident study of, 88
Trucks, study of, 24
t-test, 121–127, 129–130, 160–162
Type I and Type II errors, 126

User opinions, 24–26

Van Voorhis, W. R., 308
Variability, as a factor in increasing
 sensitivity of experiments, 163–165
 as a function of experimental controls,
 163–165
 between subjects, 236–244
 in threshold measurements, 258–259
Variability, measures of, 108–113
 percentiles, 112–113
 range, 108
 standard deviation, 108–112
 variance, 112
Variables, dependent, 215–220
 independent, 208–215
 selecting values for test, 209–213
 to be held constant, 220–234
Variance, analysis of, 127–142, 167–176,
 187–198, 300–301
 definition of, 112
 use of in correlation analysis, 117–118
Vasilas, J. N., 86–88, 309
Vernon, M. D., 259–260, 305
Verplanck, W. S., 309
Vision, dark adaptation, experiment on,
 177–179
 study of effects of anoxia on, 131–136
Volunteer subjects, 18, 158, 246
Von Felsinger, J. M., 307
von Neumann, J., 11–12, 309

Walking, index of, 59–60
Warning devices, study of, 165–176
Warren, A. B., 307
Weldon, R. J., 162, 205, 309
Williams, S. B., 248, 261, 309
Wilson, E. B., 3, 309
Woodson, W. E., 2, 309
Wording, of instructions, 225–229
 of questions, 25–26
Words for articulation tests, 286–288
Work sampling, 26–39

Yates, F., 277, 305, 309
Youtz, R. P., 86–88, 309